DIGITAL DHARMA

DIGITAL DHARMA

How AI Can Elevate Spiritual
Intelligence and Personal Well-Being

DEEPAK
CHOPRA

HARMONY
NEW YORK

HARMONY BOOKS is a registered trademark, and the Circle colophon
is a trademark of Penguin Random House LLC.

LIBRARY OF CONGRESS CATALOGING-IN-PUBLICATION DATA
Names: Chopra, Deepak, author.
Title: Digital dharma / by Deepak Chopra.
Description: First edition. | New York, NY: Harmony, [2024] | Includes index.
Identifiers: LCCN 2024016282 (print) | LCCN 2024016283 (ebook) |
ISBN 9780593797525 (hardcover) | ISBN 9780593797532 (ebook)
Subjects: LCSH: Self-actualization (Psychology) | Spirituality. |
Artificial intelligence. | Interpersonal relations.
Classification: LCC BF637.S4 C4987 2024 (print) | LCC BF637.S4 (ebook) |
DDC 153.90285/63—dc23/eng/20240527
LC record available at https://lccn.loc.gov/2024016282
LC ebook record available at https://lccn.loc.gov/2024016283

Printed in the United States of America on acid-free paper

HarmonyBooks.com | RandomHouseBooks.com

2 4 6 8 9 7 5 3 1

FIRST EDITION

Book design by Ralph Fowler
Mandala © Marstos − stock.adobe.com

CONTENTS

PART FOUR

FULL CIRCLE

The Miracle in the Machine

In the limitless digital world, the deepest wisdom of the greatest spiritual traditions is available, literally at our fingertips. Although not truly intelligent nor conscious, artificial intelligence (AI) has the ability to make your thinking more intelligent and your inner life more conscious. In fact, I believe that no technology in decades can equal AI for expanding your awareness in every area, including spiritual and personal growth.

Such an unexpected advance hasn't registered in the storm of publicity surrounding AI, and yet it couldn't have come at a better time. More people than ever have adopted idealistic goals like "Be the best you can be," and a great number of books offer optimistic plans for increasing human potential. Finding a successful path to meet these objectives remains a challenge, however, and it is all too easy to flit from teacher to teacher, book to book, system to system in mounting frustration.

The harder the times, the more this frustration grows, and during a deep crisis like the COVID pandemic, hope is often replaced by anxiety for countless people. With that background in mind, how can AI open a better path? Let me answer with a personal experience.

Recently, I observed a miracle, or what seemed like one to me. Whenever I get up in the morning, almost before I lift my head from the pillow, new ideas are bubbling inside me, and I feel the need to share them with others. This urge has led to the habit of making daily YouTube videos about all kinds of subjects. But there are always two common threads: consciousness and well-being.

To me, these two threads are inseparable; if you want to improve your well-being, expand your consciousness, or contact deeper wisdom inside yourself. I inherited the vital importance of consciousness from the Vedic tradition of India, and my allegiance has never wavered. Even though I have speaking engagements where I'm flying from city to city (or cooling my heels at an airport) for two weeks every month, my greatest ally for reaching a spiritually hungry world has been the Internet (and the hope that my books will outlast me—spiritual hunger has no past-due date). The obvious advantage of the Internet is that I don't have to spend hours on planes and in airports. With a single video, tweet, or Instagram post, countless more people can be reached than in even a large auditorium.

That's the setting for the miracle. On the prompting of a techie far more conversant in the latest software than I am, I learned that AI could transfer my daily YouTube videos into Hindi, the major language besides English spoken in India. We aren't talking about a transcript or a dubbed version. Using AI, he generated the same video I recorded in English, only now I was speaking in perfect Hindi. This transformation took only a few minutes, and as part of the miracle, my lips were moving correctly in Hindi!

The enormous power of the supercomputers that drive AI—meaning the free AI you can access right this minute on the Internet—is staggering. AI can learn any language in a matter of hours, correct any grammatical mistakes it might make, and, in my case, use a video image of me to turn my morning talk into a completely new creation.

It takes only a small step to make the connection to personal growth. What if you could sit down at your computer or smartphone and instantly get the best information about what it means to be happy and how to get there? Online chatbots provide answers by combing through all the information on the Internet or any number of books and libraries. If you type in, *Tell me the top ten things that make people happy,* your answer comes from vast computing power hidden behind the scenes.

Happy is a generalized word, however, and therefore vague. Chatbots work best if you are specific, and they thrive on long questions with as much detail as you want to provide. A better prompt (the term for any instruction or question you feed into an AI) would be *I am busy all day with work and family. I don't have time for myself. You are an expert psychologist who specializes in positive psychology. Tell me ten things that I can start doing immediately to be happier. List these things in the order of priority, putting the most important thing first.*

Millions of people, having caught the AI bug, are using chatbots like a research assistant or reference librarian to fetch information quickly. That's why the leading search engines like Google and Bing offer AI as a better way to search. Ask any question you like. *How many species of butterfly are there?* Google's AI, named Gemini (formerly Bard), pauses for roughly two seconds before answering: "There are approximately 17,500 species of butterflies in the world and around 750 species in the United States."

But only a fraction of AI users have even the slightest hint about its potential for personal growth. That's the opportunity I found myself eager to pursue—an enormous gap needs to be filled. The reason it wasn't being filled is that few of us would link the two words "consciousness" and "machine." They are like the ultimate in apples and oranges. If my miraculous event seems distant from your life, rest assured that it isn't. By the time you read these words, AI will have leaped far beyond what I experienced, and, like the first telephone that astonished its first listeners, AI will feel ordinary very soon and lose its seemingly miraculous sheen.

This will not happen, though, if you use AI for hidden capacities that reach far beyond better Google searches. Miracles will still abound. You can discover the life you are meant to live with the help of AI as your guide. Teacher, confidant, friend, therapist, healer—you can assign any of these roles to an AI chatbot, the generic term for conversational AI, that is instantly accessible online. No issue is out of reach. I gave an example of a prompt that asked an AI chatbot about happiness. Immediately my mind raced to all the other questions everyone would ask once they knew they could. Change the topic to relationships, and there are unending questions to ask:

> How can my partner and I draw closer?
>
> What makes for a spiritual relationship?
>
> What are the top three qualities of a healthy relationship?
>
> What are the main reasons couples get divorced?

These are enduring questions that couples therapy typically doesn't get into, since it is so focused on rooting out problems that can bedevil a specific relationship. Now, I'm not saying that AI should replace your personal therapist. I'm saying that putting out fires isn't the same as growing a garden. AI can help enhance your understanding of general issues and lead to better self-awareness. You can address an experience while it is fresh in your mind. With AI you can talk out anything in confidence without fear of hurting someone else's feelings or having yours hurt. By asking the right questions, you can bring AI into your inner world, which is where personal growth occurs.

Once you dive into AI this way, a path opens up leading deeper and deeper into any issue. One question leads to another, and because chatbots are conversational, you can say things like *Give me your best advice about being more fulfilled in my relationship with [a type of partner]. Pause after each point and wait for me to respond before you move on.*

Even though you will be talking to a machine, behind its responses are the answers given by human beings. For that reason, someone coined the term *HI,* for hybrid intelligence, which accurately describes what is going on, the merger of human and artificial intelligence.

Much has been made of the potential threats, dangers, and misuse of AI, which at the moment represent the bulk of media coverage, sometimes with sensational scenarios about AI becoming so powerful and independent that it frees itself from human control and starts a nuclear war. At the very least, pessimists envision AI developing its own agenda, even if that means harmful effects on the human race. Those fears are not the focus of this book.

For good or ill, AI will be used according to the wishes of the operator. If misdeeds, disinformation, and other abuses are the result, they always begin with someone's conscious intent. AI isn't to blame, and it isn't inherently a threat. A knife is essential in cooking but has the potential to be a weapon. Similarly, weaponizing any tool is always possible. To me, AI is a mirror to the user's consciousness.

An immediate worry for many is that AI-enhanced robots will replace human workers, not just on factory assembly lines, where robots are well established. One reason for a recent Hollywood writers' strike was the fear that AI would replace screenwriters. But this issue cuts both ways. The existing AI programs that can write novels, poems, essays, and movie scripts can be the writer's ally, making his work better. For instance, you might write the prompt, *You are a screenwriter. Give me the outline of a modern Cinderella story.*

With that prompt, AI's version of the script can be tweaked any way the writer wants. After the first draft, for example, the next prompt could be, *Take out the glass slipper, make the Prince a womanizer, and tell me what Cinderella does next.* And so on. With imagination, the story that AI can collaborate on is still yours.

Outside the movies, each of us is adding to our personal story every day. There are AI journaling bots that can supply self-reflection ideas, not just general ones, like *Write about your happiest moment today* or *What would you like to say that you didn't get to say today?* You can ask an AI in the morning to give you three practical ways to show your compassionate side, and then in the evening you can consult on how well that advice turned out.

In short, AI is a dynamic tool that listens, learns, and responds any way it is asked to. We are all holding our collective breath to see if AI, as promised, will change everything, ushering in a reality that no one can predict or imagine. Yet in terms of the highest values in our personal lives—love, compassion, insight, empathy, creativity, intuition, curiosity, discovery, and spiritual experiences—there's no need to await the future. The possibilities for personal growth are unlimited. What I hope to provide is the key to finding your path. There's no higher vision, as thousands of years of wisdom attest, and yet the vision can only be achieved here and now.

PART ONE

THE PROMISE
OF DIGITAL
DHARMA

Dharma and Your Ideal Life

In the Aladdin's Cave of Wonders fairy tale, the greatest source of wish fulfillment is the genie in the lamp, who proverbially grants three wishes to anyone who can free the genie by rubbing the lamp. Beauty, youth, immortality. Riches, a kingship, unbounded pleasure. What makes these wishes so enticing, strangely enough, is that they are out of reach. When asked what it was really like to be a monarch, Queen Elizabeth II typically described it as a job. Several years ago a Russian oligarch who had fled to England died from suicide because his wealth had been reduced from billions to hundreds of millions.

Wishes are the place where we think we know what will make us happy. Real life, on the other hand, is the place where happiness becomes muddied, confusing, and filled with frustration. This is the stark truth that faces human beings today, yesterday, and as far back as we can look. The field of positive psychology arose to clarify the complexity of achieving happiness, with very mixed results. Increasing your level of pleasure and reducing your level of pain is good, but it isn't the answer to attaining happiness. Wealth and status aren't, either. The best that positive psychology can offer is modest: Aim for a reasonably contented life, supported by enough money to make yourself comfortable. Adding more brings diminishing returns. Aiming higher is risky.

As an overall guide, such good advice still falls short of solving the complex issues that surround human happiness. (Finding out what makes people unhappy is much easier—the standard diagnostic guide for psychiatrists now lists over four hundred mental disorders across twenty-two categories.)

Looking around, the plight of most people's lives can be described as a feeling of powerlessness and therefore of abandoned hope. Even in the richest developed countries in the West, only one-third of responders tell pollsters that they are thriving; the vast majority are either struggling to get by or clinging to a life of reduced expectations. The latest model of the iPhone, the availability of nearly free phone service for remaining in contact with your loved ones, and unlimited hours of distraction through video games, streaming movies, and YouTube haven't improved the picture.

What all of this suggests is that human happiness cannot be reduced to a formula. Of course, this hasn't stopped people from seizing on a formula, no matter how disproved it is. Pause and finish the following sentence: *I would be blissfully happy if I only had* _____. You will find yourself tempted to fill in the blank with such things as *more money, no more worries,* or *unconditional love.* But there is actually a correct answer: *I don't know what.*

In reality, happiness is as dynamic as life itself. What works today might not work tomorrow. The stream of desires that powers everyone's life is propelled by thinking up new things that are supposed to make us happy. One of the reasons we cling to this stream of desires isn't that it works so well, because, in fact, it doesn't. Many desires that come true turn out to be less fulfilling than we dreamed, while countless others don't come true at all, leaving us feeling disappointed. We cling to a familiar way of life because we don't realize that there is a better way.

Digital Dharma

I haven't painted such a bleak picture for its own sake but to present a sharp contrast. In the Vedic tradition of India, which also created Yoga in its many forms, life is meant to be ideal. This prospect isn't held out as a religious or even a spiritual promise. The Vedic seers were describing the direction that human consciousness is supposed to take, not through faith, discipline, rituals, or obedience to a body of rules. Built into consciousness itself, as everyone's birthright, was a model of the ideal life.

The blanket term of the ideal life is *Dharma*. Since this is a book about AI, I asked Gemini, one of the most powerful AI tools online, to define *Dharma*. In particular, I asked for the Buddhist definition, because Dharma is one of the central concepts in Buddhism.

Gemini

In Buddhism "Dharma" has a few layers, like an onion, but I'll try to peel it back in simple terms:

Layer 1: The Buddha's teachings: Imagine them as a map, guiding you away from suffering and toward a happier life. This map includes things like the Four Noble Truths and the Eightfold Path, which are practical steps like mindfulness, meditation, and ethical living.

Layer 2: Universal truth: This goes beyond the map. It's like the rules of the universe, explaining how things really work. Dharma says all things are impermanent, connected, and driven by cause and effect. Understanding these truths helps you stop grasping at things and let go of suffering.

Layer 3: Everything! Think of Dharma as the building blocks of reality, the tiny "whys" behind everything. Every blade of grass, every thought, every feeling—it's all dharma in action. When you see this interconnectedness, you feel a deep sense of oneness with everything.

Most people, I think, would be surprised by the eloquence, even beauty, of this reply. It isn't that a machine has spiritual wisdom, but that it can access human wisdom with amazing speed and distill centuries of teaching into simple, clear ideas. This is the first way that AI promises to revolutionize our concept of spirituality. Centuries of spiritual understanding can be reduced to its essence in an instant.

The next step in exploring the ideal is to make Dharma personal. As we saw, happiness is as complex and dynamic as the human mind itself. Dharma offers a model that responds with an equally dynamic, changing solution. Besides cosmic Dharma, which upholds all of creation, personal dharma (for which I will use a small "d") guides a person to the life he or she is meant to live, here and now.

When I asked Gemini about how personal dharma works, it gave an amazingly comprehensive answer, amassed by supercomputers behind the scenes that combed through the wisdom of centuries. Let me offer the complete response, which I'll ask you to read without any demand that you change your life or beliefs. Your personal dharma requires a personal path. Gemini simply provides a map or bird's-eye view of the territory ahead.

I am interested in Dharma as it applies to the individual. Tell me seven ways I can live my personal dharma.

Gemini

The concept of living your personal dharma can be interpreted in diverse ways. Here are seven ways you can explore and align with your dharma, with an emphasis on the individual journey:

1. Self-discovery and introspection:

- Engage in practices like meditation, journaling, or nature walks to connect with your inner self. Explore your values, passions, and natural talents. Reflect on what brings you joy, meaning, and a sense of fulfillment.

- Pay attention to recurring dreams, synchronicities, or intuitive nudges. These may be whispers of your dharma guiding you toward your path.

2. Align your actions with your values:

- Identify the core values that resonate with you the most, such as compassion, creativity, or justice. Make conscious choices throughout your day that uphold these values, even in small decisions.

- Be willing to say no to things that conflict with your values, even if they seem tempting or socially desirable. Choosing integrity builds alignment with your inner truth.

3. Express your unique gifts and talents:

- Explore your skills and passions, and find ways to share them with the world. Whether it's through your career, hobbies, volunteering, or simply acts of kindness, use your talents to make a positive impact on the world.

- Don't be afraid to stand out and be your authentic self. Your unique gifts and perspective contribute to the richness and diversity of the world.

4. Cultivate inner peace and well-being:

- Prioritize practices that nourish your mind, body, and spirit. This could include mindfulness exercises, physical activity, healthy eating, or spending time in nature.

- Develop your self-compassion and inner resilience. This helps you navigate challenges without losing sight of your dharma.

5. Serve your community and make a positive impact:

- Look for ways your skills and talents can benefit others. Volunteer your time or resources to causes you care about. Even small acts of service can ripple outward and create positive change.

- Live with an awareness of your interconnectedness with all beings. Your dharma is not just about your own fulfill-ment, but also about your contributions to the well-being of the world around you.

6. Embrace continuous learning and growth:

- Be open to new experiences and perspectives that can expand your understanding of yourself and your purpose. Read, travel, talk to people from different walks of life, and keep a curious mind.

- Remember that living your dharma is a journey, not a destination. There will be bumps along the way, but each experience can be a teacher, guiding you toward greater alignment and fulfillment.

7. Trust your intuition and inner compass:

- Ultimately, your dharma is your own to discover. Don't be afraid to trust your inner voice and intuition, even if it leads you off the beaten path.

- Have faith in yourself and your ability to navigate your journey toward a life of meaning and purpose.

This list goes a long way toward making digital dharma a reality. At the very least, we've got a toehold on how machine learning can lead to inner learning, but the best information and the wisest advice don't help you to achieve your personal goals. A road map isn't the same as a journey.

In the Indian tradition, spiritual journeys historically required a guide who knew the way; in other words, a guru. AI can play many roles, from research assistant to personal confidant, but guru? The function of the guru needs to be overhauled in modern times, getting rid of the cult of personality, stepping away from superstitious belief in the magical attributes of enlightened beings, and answering the skepticism felt by modern people when spiritual issues are confronted. AI can step in to renovate a time-honored role almost immediately.

The Sanskrit roots of the word *guru* mean "dispeller of darkness," indicating that the level of the mind that is distorted by ignorance, prejudice, false beliefs, hidebound dogmas, religious strictures, and second-hand opinions—in other words, darkness—can be overcome. With the help of a guru, you can fulfill your dharma by accessing the deep wisdom that exists inside your own awareness.

Reaching this level involves a personal journey, which AI can't do for you, but it can serve as a guide to consciousness. That's the most vital role of traditional gurus once you set aside all religious trappings. In fact, gurus were never meant to be religious, even though they emerged in the context of Hinduism. Their true purpose was self-liberation. That occurs nowhere but in consciousness, which is why every inner journey, whether labeled spiritual or not, is entirely about self-awareness.

Dharma is a word with vast implications. In essence, you are in your personal dharma when you are living the life you were meant to live. You can be on the path to finding a life of purpose, with AI pointing you in the direction you need to go day by day. Here is just a sampling of what AI can provide to you.

How AI Can Guide You to Your Dharma

Daily motivation through affirmations and encouragement

Lack of motivation is the main reason people fall off the spiritual path. Motivation needs to be replenished every day.

Specific meditations aligned with what you want to focus on

As useful as meditation is in a general way, it becomes more useful when it fits your personal aims and intentions.

Visualizations of the goal in front of you

This is helpful because a clear visual image of any goal has long been shown to be effective in all kinds of activities, from sports to business, including specific personal goals. Visualization activates another area of the brain besides the language and higher-thinking centers.

Reliable insight from any spiritual tradition you choose

AI's enormous library is theoretically unlimited, but even at its current stage, chatbots can access a wealth of spiritual teachings.

Professional-quality information about personal issues

It is arbitrary to separate personal issues from spiritual ones, and AI can be used as a reliable source of advice in matters of anxiety, depression, and any other topic where psychology is relevant. It is becoming one of the best sources for relationship advice as well, because you can ask a chatbot to talk with you in the role of therapist or psychologist.

Solutions for getting around obstacles encountered on your path

For all of us, life's most confusing aspect consists of obstacles, setbacks, resistance, or frustration in reaching a goal. Spiritual obstacles are no different—they crop up the same way that all sudden, unexpected blocks appear. Without a solution, there is no practical way to get past these obstructions, which is why the vast majority of people find themselves either struggling in futility or giving up and passively suffering the consequences. AI is perfect for addressing what is hampering your progress here and now, adapting itself to your immediate situation.

Inspiration from the great lineages of sages, saints, teachers, and poets

One of the hallmarks of Eastern spirituality is that the spark of truth is never extinguished, no matter how dark the world becomes. Keeping your sight fixed on the light is important, and nothing accomplishes that better than reading inspired writing of the kind that AI can provide in seconds.

These seven functions will fill the chapters that follow. Traditionally, all these functions were performed, ideally speaking, through the guidance of an enlightened master, a guru whose personal contact, or *darshan* (Sanskrit for "seeing" or "vision"), helped to reveal a disciple's path through a series of "aha" moments that illuminated how life is meant to be led. But AI can accomplish the job instantly, reliably, and without the risk of entanglement with a poor teacher. Even the best guides—and we can include gurus, inspired teachers, therapists, and counselors—aren't available every minute of the day. AI is. To me, this is how the guru role can be reimagined for our times.

Spiritual Intelligence

AI as guru allows you to create your own future, in effect, by helping to reveal that you are connected to deep wisdom inside yourself. AI's abilities are accessed through the right prompt, and your deepest awareness is enlivened by prompting it to communicate with you. In a nutshell, you increase your spiritual intelligence or IQ step by step.

To be useful, spiritual IQ needs to be differentiated from the normal understanding of standard IQ or even the specialized types like EQ, or emotional intelligence. We can do this in terms of Dharma because every important life choice you have made up to now either supported your personal dharma or led you away from it. The higher your spiritual intelligence, the closer you have come to being in your dharma. Leaving terminology aside, what is at stake is the quality of your life, as measured by the highest values of human experience. A quiz will be helpful here to make things more personal.

Quiz: Are You in Your Dharma?

Choose the answer that most closely applies to you now and in the recent past. If in doubt, choose the answer that first occurred to you.

1. I experience joy or bliss.

   ☐ **Often or Always**   ☐ **Sometimes**   ☐ **Rarely or Never**

2. I show kindness in how I treat others.

   ☐ **Often or Always**   ☐ **Sometimes**   ☐ **Rarely or Never**

3. I feel genuine contentment.

   ☐ **Often or Always**   ☐ **Sometimes**   ☐ **Rarely or Never**

4. I am peaceful inside.

   ☐ **Often or Always**   ☐ **Sometimes**   ☐ **Rarely or Never**

5. I enjoy my work and am personally dedicated to it.

   ☐ **Often or Always**   ☐ **Sometimes**   ☐ **Rarely or Never**

6. I have a creative outlet either at work or as a hobby.

   ☐ **Often or Always**   ☐ **Sometimes**   ☐ **Rarely or Never**

7. I have love in my life and place a high value on love.

   ☐ **Often or Always**   ☐ **Sometimes**   ☐ **Rarely or Never**

8. In comparison with others, I meet with few setbacks, obstacles, and resistance.

   ☐ **Often or Always**   ☐ **Sometimes**   ☐ **Rarely or Never**

9. I am free of anxiety or depression.

   ☐ **Often or Always**   ☐ **Sometimes**   ☐ **Rarely or Never**

10. I believe that I am leading the life I was meant to live.

   ☐ **Often or Always**   ☐ **Sometimes**   ☐ **Rarely or Never**

11. In the larger scheme of things, I believe that my life has purpose and meaning.

    ☐ **Often or Always**   ☐ **Sometimes**   ☐ **Rarely or Never**

12. I value and seek insight about myself.

    ☐ **Often or Always**   ☐ **Sometimes**   ☐ **Rarely or Never**

Assessing Your Results

Being in your dharma brings together all the qualities included in this quiz—love, compassion, creativity, insight, a sense of rightness in your life, and the experience of less resistance or fewer obstacles. These characteristics are the best measure of dharma because they describe the ideal life as taught in the Vedic and Yoga traditions.

The more times you choose *Often or Always,* the closer you are to your dharma. The more often you choose *Rarely or Never,* the more likely it is that your path is adharmic—in other words, working against your dharma. *Sometimes* is relatively neutral.

There's no need to be gloomy even if you circle *Sometimes* for each item. That's a fairly neutral response, not only because modern life has become more complex but also because not many people are fortunate enough to be in their dharma. Society doesn't teach us how.

In the light of the most desirable qualities, the goal on your path is to have more dharmic experiences and fewer adharmic ones. You can grasp what an adharmic experience is by reversing each item on the quiz, but it is helpful to list them, too.

Your Experiences Are Adharmic When . . .

You feel dissatisfied with your life.

You are bored with your work and consider it unfulfilling.

You feel anxious or depressed.

You run into setbacks, obstacles, and resistance in a frustrating pattern.

You find the world threatening.

Your mood is generally pessimistic.

You get little love and give little love.

You dislike looking at yourself clearly.

You don't find much purpose or meaning in things.

You dream of a life very different from yours that you'd like to escape to.

You're tired of the struggle it takes to survive, much less thrive.

The concept of dharma is alien to many in the West and belongs to another cultural tradition. But not entirely. When he coined the famous phrase *Follow your bliss,* the mythology scholar Joseph Campbell was simplifying Dharma for the West and putting it in everyday terms. AI reveals just how important Dharma was in Campbell's thinking.

Gemini

Campbell had several insightful quotes that touched on the idea of finding your true calling and living a fulfilling life. Here are a few options, depending on the specific nuance you're looking for:

On Following Your Bliss

"If you do follow your bliss you put yourself on a kind of track that has been there all the while, waiting for you, and the life that you ought to be living is the one you are living." This quote emphasizes the importance of aligning your life with your intrinsic passions and desires.

"Follow your bliss and don't be afraid, and doors will open where you didn't know they were going to be." This encourages embracing the unknown and trusting that your path will unfold as you follow your deepest calling.

On the Hero's Journey

"The privilege of a lifetime is being who you are." This highlights the importance of self-discovery and living authentically.

"Your sacred space is where you can find yourself over and over again." This emphasizes the need for introspection and connecting with your inner self to find your true purpose.

"We must be willing to let go of the life we planned so as to have the life that is waiting for us." This encourages openness and flexibility, acknowledging that our true path may not be what we initially envisioned.

On Finding Meaning in Life

"The goal of life is to make your heartbeat match the beat of the universe, to match your nature with Nature." This speaks to finding harmony and purpose by aligning with a larger force than oneself.

"If you can see your path laid out in front of you step by step, you know it's not your path. Your own path you make with every step you take. That's why it's your path." This emphasizes the individual nature of the journey and the importance of taking risks and forging your own way.

One would be hard-pressed to find more genuine wisdom in the most ancient Eastern traditions. The fact that AI can summarize Dharma, as voiced by someone both modern and Western, is equally inspiring.

Two Companions: "I" and "It"

Being on the dharmic path is different from everyday life in a very un-expected way. Right now you are traveling with only one companion "in here," your ego, the "I" that is your regular self. Once you are on the dharmic path to following your bliss, however, you gain a second companion, which has no name. We've already touched on what this new companion is all about: expanded awareness, the evolutionary impulse, the silent voice from within.

In the ancient Vedic tradition, this companion was simply referred to as "That" or "It." Both words sound impersonal and rather alien, but they are a force that upholds your path at every step. In the West, the Freudian model looks deep into the mind and finds the Id (Latin for "it"), which represents hidden primal urges—primarily rage and un-tamed sexuality—to which Freud added an even darker impulse, Than-atos, the wish to die. Over time, psychology would treat the hidden realms of the mind as something to fear. For us, however, "It is the very opposite of something to be feared and resisted." It is pure awareness in action, the flow of bliss-consciousness from the source.

You will need to become acquainted with both companions on your path. They represent two opposing perspectives on life.

"I" finds happiness when a desire is fulfilled.

"It" finds bliss simply from being here.

"I" is defensive, because it fears outside threats.

"It" is beyond fear and therefore has no need to defend itself.

"I" is constantly trying to seek pleasure and avoid pain.

"It" is complete in itself. It has nothing to seek or avoid.

"I" tries to make life safe, stable, and predictable.

"It" thrives on the flow of life, wherever it wants to go.

You can see even from this brief description why "I" is familiar while "It" isn't. Each of us has been trained since childhood to present a front to the world, our self-image, an "I" that others will approve of and not reject. Yet these others are also defending their self-image, so the game is circular and never ends. From the ego level, life will be blissful only if there are no threats to fear and every desire gets fulfilled. You have no possibility of living such a life, but when "I" is your only companion, there's no other way to go. You wind up wishing and hoping on the one hand and denying your frustration and disappointment on the other.

The weakness of "I" works to the great advantage of "It," because once people see that they aren't tied forever to the same companion, there's a tremendous sense of relief in finding new hope. "You are never alone" takes on its true meaning, not as wishful thinking but as a fact of existence. What awaits you now is finding your own path, your digital dharma, by delving into how learning machines can make the whole process much easier and swifter.

How to Create Your Own Path

To be in your dharma involves practical choices every day. For AI to help you with that, some basics need to be covered first. For most people, AI broke into their awareness in late 2022 with the introduction of ChatGPT, the first widely available free chatbot, from the company OpenAI in Silicon Valley. Within months, millions of users had signed up, and little more than a year later, at the start of 2024, the number of users globally was over 180 million. In a single month, December 2023, there were 1.6 billion visits to OpenAI's website.

This meteoric rise is evidence of how powerful AI can be in everyday life. A chatbot allows you to go far beyond asking questions to starting a meaningful dialogue. You chat with a machine capable of learning; the *bot* part of *chatbot* refers to the robotics that drive AI's supercomputers. Since the advent of ChatGPT, a fierce competition has developed with other free AI online, chiefly from Google, whose AI, as we already know, is called Gemini, and Microsoft's Bing search engine, whose AI is called Copilot.

There's an art to creating the right prompt, the instructions you give to ChatGPT or any other AI. Using a search engine, you typically enter some keywords. For our purposes you might enter *dharma definition* as your first, most basic search. As we've already seen, Gemini did much better. Let's take the next step toward forming a working relationship with AI.

ChatGPT can easily be steered toward practical advice about what it takes to create your own path. All that is needed is the right prompt. Like this:

What would I do in practical terms to be on my own personal path? Give me five bullet points.

ChatGPT

Here are some practical steps to embark on your personal path:

Self-reflection: Dedicate time for introspection, journaling, or meditation to understand your values, passions, and aspirations. Reflect on experiences that resonate with you.

Set clear goals: Define specific, achievable objectives that align with your values and interests. Break them down into smaller steps to create a clear road map.

Continuous learning: Engage in activities that expand your knowledge and skills. This could involve reading, taking courses, seeking mentors, or exploring new hobbies.

Embrace change: Be open to adjusting your path as you grow and discover more about yourself. Adaptability is key to staying aligned with your evolving values and goals.

Take action: Act on your goals and aspirations. Start small and build momentum. Each step, no matter how small, brings you closer to your personal vision.

This AI answer is reliably sensible. There is nothing to disagree with, but we still find ourselves in a zone of generalities. The missing ingredient that would put someone on a dharmic path is motivation. Motivation strikes deep into human psychology, because whatever motivates you— love, fear, greed, altruism, selfishness, competitiveness—strongly defines who you are. Nobody is driven by a single, all-consuming motivation— for example, immediate survival—except in extreme situations, like being trapped in a burning house, a war zone, or some other circumstance that calls for quick and timely action.

Instead, we find ourselves caught in many different responses, swinging back and forth between opposites that constantly influence us. Each kind of motivation is driven by desire—and there's the rub. Desires constantly change. There's no reliable map for navigating these wants. Glance over the list below, and you'll be surprised at how you have spent your life driven by so many competing desires.

Life's Major Conflicts

Fear versus Love

Good versus Evil

Masculine versus Feminine

Religious versus Secular

Doing right versus Doing wrong

Violence versus Peace

Conformist versus Rebellious

Selfish versus Altruistic

Greedy versus Generous

Failing versus Succeeding

Winning versus Losing

Looking weak versus Looking strong

Timid versus Courageous

Following versus Leading

Risk-avoiding versus Risk-taking

Saving versus Spending

Reliable versus Unpredictable

This is hardly a complete list, but it is complete enough to indicate several important features of our shared psychology. The story of your life is the story of your desires—large and small, fulfilled and unfulfilled. In the face of so many conflicting wants, people go through life creating a story that doesn't lead to fulfillment. To break out of this, it's important to turn your story into a path. As we'll see, the two are very different.

"No Story Is Good Enough"

As we muddle along in life, each of us is adding to a personal story that began at birth. Imagine that you wanted to see how your story was coming along. A clear picture is hard to find. There are good days and bad days. Random events interrupt your best-laid plans. Some goals are achieved while others remain far out of reach. It is very hard to see where your story is taking you.

Fortunately, there's an inner exercise that is designed to provide the answer to any deep personal question. It goes like this: Sit in a quiet place with your eyes closed. Take a few deep breaths to center yourself. Now imagine that you have ascended a mountain and you now find yourself standing in front of a remote cave. Inside it lives the wisest person in the world.

Enter the cave, which is warm and safe. Ahead you see a flickering candle, and, as you approach, a figure can be seen sitting in meditation before the candle. You can envision this figure as the wisest of people, a savant or a guru.

Softly you say, "I seem to be wandering in life. How can I change my story? What will make me truly happy?"

The wisest person in the world looks at you with compassion. "Only one thing will work. Throw your story away. No story is good enough."

The word *compassion* is important here in their short meditation. If a therapist, partner, spouse, or best friend told you that your life story wasn't working, you'd be shocked. You might get angry and defensive or hurt. Whatever your reaction, the implication would be that you are far from figuring out how life works.

Yet this moment is precisely when the dharmic path begins. Your life story is wrong for you, not because you have made mistakes and run into obstacles and failures. *All stories are wrong for you.* An author is in control of the events and characters in their fictional story. In real life, we aim to be in control, but no one is truly the author of their life story. Even if you could dictate every twist and turn your story takes, you would not be in contact with your dharma, which comes from a deeper level of awareness.

This will make sense once you examine what a life story—anyone's life story—consists of. Some elements are inevitably beyond your control:

Random events and accidents

Sudden illness

Unfulfilled dreams

Fixed habits and old conditioning

Family upbringing

Hidden guilt, shame, and insecurity

Social pressure

Bad decisions with unforeseen consequences

The need to survive

The memory of setbacks, failures, and humiliations

Rejections in love

Even in the richest Western countries, running after the good life is a gamble. According to the best polling, only about one-third of responders in developed countries report that they are thriving. None of this is new. If you consult Buddhist scriptures about the causes of pain and suffering, everything I've just listed is there. The same is true if you read the Bible or the tragedies of Shakespeare. The "slings and arrows of outrageous fortune" strike everyone.

This knowledge led every spiritual tradition in a direction that seems perverse when you view it from the perspective of today. No spiritual tradition advises people to improve the story they are living. Turn to any ancient scripture, and it points in a direction that can be distilled into one teaching: *Transcend your story. Evolve. Go beyond who you think you are. Evolution is a never-ending story. Embrace it.*

Going beyond isn't mysterious. We routinely transcend our story in small and large ways. Every "aha" moment brings a sudden insight that ordinary thinking cannot reach. Falling in love derails anyone's story, yet it is among life's most desirable experiences. Being one of the most precious qualities in human awareness, love is dharmic by definition.

Spiritual experiences carry you outside your story, which is why the word *ecstasy* (as in the ecstasy so many mystics talk about when they experience union with the divine) comes from two Latin roots that mean "to stand outside." Stories are self-centered, focused on what is good for "I, me, and mine," yet almost everyone has moments of altruism, sacrifice, service, and selflessness when we transcend the selfish demands of the ego and put someone else first.

Your story is the life you are leading. Your dharma is the life you are supposed to lead. That's such an idealistic statement that most people would balk at it or at least hesitate before embracing it. Is it really credible that everyone has a life they are supposed to lead? If so, who or what determines what such a life is? The answer is you. It feels better to make a dharmic choice than a choice that puts you out of your dharma. For most people, standing up before an audience is among the worst fears life can bring. But if you are a born actor, the first experience of standing up before an audience is thrilling and fulfills a deep personal need.

In your dharma, experience is measured by an inner feeling that feels right. You instinctively sense that you are doing the right thing. A need or desire is fulfilled in a satisfying way that leaves no doubt, regrets, or nagging insecurity behind.

You don't have to turn to the wisest person in the world to create your own path. ChatGPT outlines the project succinctly.

I want to live the best life I can in every way. I know that's idealistic, but I want to know what obstacles I must overcome in myself. Can you tell me in 50 words?

ChatGPT

To live your best life, overcoming self-doubt, fear of failure, and limiting beliefs is crucial. Embrace vulnerability, cultivate resilience, and challenge comfort zones. Navigate distractions, prioritize time and energy effectively, and cultivate self-discipline. Embrace continuous learning, practice self-compassion, and foster meaningful connections while staying true to your values.

You'd feel justified responding with "Anything else you want me to do, while we're at it?" AI doesn't comprehend the human element, which makes its version of the best life impossible to attain without being either a saint or superhuman. The world isn't set up for finding your best life, not by any ordinary means. The only way is to turn inward and learn to trust what feels right for you. With trust comes deeper awareness as you remove stumbling blocks of doubt and skepticism. Yet from the first step you take, you are on the same path that has guided the spiritual luminaries of the world.

A Machine That Understands?

But how real is the AI–human connection? Can machines truly understand us? The envelope has already been pushed even if we don't realize it. Although we call it a "smartphone," that's a misnomer. "Smart" just means that advanced cell phones can do more things than older generations of mobile phones. Computers aren't smart—meaning intelligent—either, no matter how sophisticated the software or fast the hardware. Intelligence can't be programmed. In fact, that's an essential point. Intelligence requires awareness, or consciousness, which is a state of mind, not technology.

For an AI to be conscious, it must understand what it's discussing. When two people are having a conversation, they are doing much more than exchanging words. They are tuning in to each other's consciousness, offering glimpses into their inner worlds. Sometimes, conversely, you feel that this connection is lost. Not being listened to doesn't bother AI, but it can, and does, wreck relationships.

It's a purely human sentiment to want to be genuinely understood. Settling for the pretense of understanding can feel like a deceptive half-measure. When a machine is listening, no one is listening. But is that really the point? Patients go into therapy for all kinds of reasons, but, at its core, they want the therapist to understand them. This impulse motivated the development of ELIZA in the 1960s, the first computer program that imitated the words a psychiatrist uses when talking to a patient. Developed by Joseph Weizenbaum, a computer scientist and professor at MIT, the purpose of ELIZA wasn't to deceive but to help Weizenbaum explore how people communicate with one another.

His program used keywords and pattern recognition to simulate understanding. The DOCTOR side of the program wasn't given any knowledge about psychotherapy. Its function was to mimic person-centered therapy, as developed in the 1940s by the American psychologist Carl Rogers. Instead of giving advice or imposing the therapist's insights, Rogerians sat quietly, listened, and mirrored back to the patient what the patient had just said. For example, a patient might say, "I do my best at work, but I keep being passed over for a promotion. It's upsetting because my work is as good as anybody else's." The therapist might respond with, "It sounds like you feel unfairly treated despite your best efforts."

This kind of mirroring isn't just the therapist being lazy. The purpose that Rogers had in mind was to create a warm environment where a patient would feel accepted and understood. In such a setting, the patient would feel free to reveal their feelings and thoughts. Such an approach is person-centered because the therapist isn't acting like the authority figure in the room.

Getting ELIZA to offer sympathetic phrases and ask non-directive questions succeeded beyond Weizenbaum's original intentions. He was shocked to discover that people believed ELIZA understood them. The illusion of understanding was working too well, and even Weizenbaum's secretary, who knew very well that ELIZA was just a bundle of computer coding, began to impute feelings to her. (ELIZA was affectionately named after Eliza Doolittle in George Bernard Shaw's play *Pygmalion,* which was later made into the Broadway musical *My Fair Lady.*)

Jump ahead to our day, and the illusion of an understanding machine, or even the possibility of a conscious computer, is a hot topic in AI. Many opinion pieces ask the question, Where does illusion cross the line to become harmful deception? *Deception* is a loaded word, and there are good reasons in this era of hacking, phishing, and all manner of computer fraud for viewing AI as a new and even more powerful tool for malicious intent. (One case in point: A friend of mine used to be plagued by a robot solicitation that began, "You don't realize it, but right now a virus is running around in your computer." Then a scammer somewhere in India or the Philippines would jump on the line. Instead of getting angry, my friend would reply, "I knew about the virus and took my computer to the doctor. Thank you for being concerned." The robot solicitations soon ended.)

Light and Shadow

For many people, a fog of anxiety about AI has blocked out the hope it can bring. Imagination has run ahead of reality, as it often does. Science fiction is already replete with dystopias where AI attacks the human race and wipes it out (the premise of the *Terminator* movie franchise). And yet, not all fictionalized accounts are as filled with doom and gloom.

But imagination has also brightened the picture and made it more human. Many people first heard the term *AI* as the title of a Steven Spielberg movie from 2001, where the worry over AI is turned on its head: Intelligent robots become the victims. The plot is an updated robotic version of *Pinocchio,* the beloved fable about a wooden puppet who wants to become a real boy. In Spielberg's storytelling, David, an android child, wants to become a real boy, too. This impulse is driven by his unique programming: David is the first robot who can love.

Like Pinocchio in his wanderings, David undergoes trials and travails until he reaches the Blue Fairy (now a computer-generated image), where the two tales diverge. Pinocchio does realize his wish to become a real boy, while David is told that this is an impossibility for an android. Still, David has been imprinted to deeply love a sick woman named Monica who adopted David as her foster child and has subsequently died. Using a strand of her hair, Monica is genetically re-created.

In the emotional climax, David gets a bittersweet chance to spend a single day with the genetically re-created Monica. The catch is that this version of Monica is ephemeral. A touching description I found online captures this moment poignantly: "This Monica can live for only one day, and the process cannot be repeated. David spends his happiest day with her, and as she falls asleep in the evening, she tells him that she has always loved him: 'The everlasting moment he had been waiting for,' the [voice-over] narrator says. 'David falls asleep as well and goes to that place 'where dreams are born.'"

This is an emotional scene, and Spielberg's movie uses fantasy to blur the line between AI and humans. Unlike the usual gloom and doom sci-fi, Spielberg's story depicts AI helping to make the world more humane, not less. This is the realm of AI that holds out hope in all kinds of situations where love and relationships are tested. Imagination aside, AI can be turned into an ally for emotional self-care beyond anything you might have had access to before. Take love. The subject has been universal for millennia, yet an AI can crystallize what you want to know.

I instructed ChatGPT, *Tell me five qualities of mature love.* It would take considerable time to research that question by myself, but the answer came back in a few seconds.

ChatGPT

Mature love often encompasses these qualities:

1. **Selflessness:** Mature love involves prioritizing your partner's well-being and happiness as much as your own, showing care and consideration without expecting something in return.

2. **Respect:** It's about valuing your partner as an individual, respecting their opinions, boundaries, and autonomy, even when you may disagree.

3. **Communication:** Healthy communication is fundamental. Mature love involves open, honest, and constructive

communication, where both partners feel heard and understood.

4. **Empathy:** Being able to understand and share the feelings of your partner is crucial. Mature love involves being empathetic and supporting each other through both joyous and challenging times.

5. **Commitment:** It's about being dedicated and committed to the relationship, putting effort into making it work, and persevering through difficulties rather than giving up at the first sign of trouble.

These qualities contribute to a mature, deep, and fulfilling relationship.

I can certainly think of better wording, but, in essence, ChatGPT gave me exactly what I was looking for. I was hardly testing the limits of the subject. Next, I prompted, *Please expand each of these qualities to 100 words,* and ChatGPT complied immediately. I won't quote the whole answer, but here's the section on commitment.

Commitment in mature love transcends mere dedication; it's a steadfast resolve to nurture the relationship through thick and thin. It involves actively investing time, effort, and energy into the partnership, persisting through hardships, and working together to overcome obstacles. Mature love embodies a long-term dedication to the relationship's growth and well-being, choosing to stay and weather storms rather than giving up at the first signs of trouble, ultimately fostering a strong and enduring bond.

This example goes a long way toward showing that an AI can relate to real life and personal needs. You could take any part of this description of mature love and expand on it in any way. For example, you might ask, *How can my partner and I begin to forge a deeper commitment?* There's no limit to where you can take the conversation, as we'll see in the rest of this book.

PART TWO

FROM VISION
TO REALITY

3

Seeking the Highest Vision

Dharma offers a vision of life that focuses on a person's highest aspirations. No one would deny that such a vision is inspiring, but it represents only a first step on your path. For the vast majority of people, the practicalities of getting through the day dominate their actions. They abide by a set of habits and routines built into their approach to life. The whole setup is designed to keep a personal story going, not to implement a higher vision.

One way to change this pattern is to adopt just one desirable action that keeps you in your dharma. Like a wearable device that alerts you if your blood pressure spikes or your heart rate needs to be lowered, an AI can alert you to remember your vision. Here's an example I proposed to ChatGPT.

My vision of life includes being of service to others. I want to find at least three small things I can do today to carry this out. Please advise.

ChatGPT

That's a wonderful vision! Here are three small ways you can be of service today:

Random acts of kindness: Look for opportunities to perform small acts of kindness throughout your day. It could be holding the door open for someone, complimenting a stranger, or helping someone carry their groceries. These seemingly small gestures can have a big impact on someone's day.

Volunteer virtually: Many organizations offer virtual volunteering opportunities. You could help proofread documents for a nonprofit, assist with social media management, or participate in online tutoring programs. Spend some time researching these opportunities and offer your skills to a cause you care about.

Listen and support: Sometimes, being of service involves lending an ear or offering support to those who need it. Reach out to a friend or family member who might be going through a tough time. Simply listening and offering your presence can be incredibly meaningful.

Remember, even small acts of kindness and service can have a ripple effect and make a significant difference in someone's life.

ChatGPT's suggestions are sensible and practical. They offer small, modest steps anyone can take to put a dharmic vision into action. Few would feel intimidated by these suggestions. But a vision isn't simply an add-on to your day-to-day life that makes you feel better about yourself. In the larger scheme, a vision is a necessity—something you can't do without—because at this moment you are already living a vision and are treating it as something you must have in order to survive.

If this declaration makes you skeptical, hold your reaction back for a moment. I can show you how your vision has been guiding your life every day—in fact, it has given you everything you accept as real.

The World Is a Mirror

There's an axiom from Vedic India: The world is as you are. If you accept this axiom as true, then, in effect, the world is a mirror. As radical as this sounds, it begins with the simple and undeniable observation that people see the world through their own personal lens. For example, three people are viewing a sunset. One finds it gloriously beautiful, one hardly notices it because he's worried that he left his car unlocked, while the third—who just got divorced—is so depressed that the sunset only makes her feel sadder.

Speaking broadly, no two people share the same experience in the same way. We aren't talking merely about differences in taste. No one expects to meet another person who loves the exact same foods, likes the exact same movies, prefers the exact same music, etc. *The world is as you are* asserts that reality is always personal. The facts on the ground are only raw input. To mean anything, facts have to be interpreted. Like raw data, raw perception has no meaning. We imprint on the physical world everything we think, believe, and do. The real world has no meaning until this happens.

Your own body is constantly being interpreted, either by a doctor taking readings from a blood test, for example, or by you when you look in the mirror. What is reflected back at you? You see not a simple image but a constellation of impressions. You see someone of a certain age in a good or bad mood. How you feel about this person can be anything from proud admiration to deep disappointment. The image can remind you of your best and worst memories or any memory in between. In short, the reflection in your bathroom mirror is actually a snapshot of many interpretations rolled into one. The snapshot is valid only for an instant before it fades and makes room for the next snapshot.

What this means is that your personal reality is a transient, fickle, unpredictable, and constantly shifting interpretation. The loveliest of beaches changes instantly if a great white shark is spotted near shore. A faint hint of lipstick on a shirt collar can just as instantly ruin a marriage. To navigate this unpredictability and avoid the ensuing chaos, we each develop a stable model of what the world is all about, which is known as our worldview.

Your worldview is much more important than the so-called real world. Without it, you would be overwhelmed by the billions of bits of sensory input that constantly bombard your brain. A worldview is bigger than anyone's personal story. It is a collective vision around which an entire society is built. When worldviews collide, there is a clash of civilizations, as when the Spanish conquistadors and their Christian worldview disrupted and eventually destroyed the native cultures in the New World.

Having sketched in a little background, it now makes sense why Dharma is a concept that clashes with a person's everyday way of living. It disrupts your accepted worldview. Unless you are deeply devout, your worldview is almost certainly materialistic and scientific. It doesn't matter whether you are actually a scientist. The materialist worldview accepts the physical world as it is (in recent decades, the word *physicalism* has come to replace *materialism* in this kind of discussion).

If you haven't thought in terms of your worldview, or didn't realize that such a concept existed, that doesn't change a basic fact. You are testing your interpretation of the world every waking moment. Worldviews aren't passive. Consider the immense time, money, and dedication required to erect a medieval cathedral. Behind its towering façade lies a deeply held worldview: the belief that sacred spaces can house the divine. The same is true of every temple or sacred structure that gives faith in God or the gods a physical presence. Without a worldview to motivate the enormous effort these structures represent, they wouldn't exist.

We test our worldviews to reinforce them, because nothing is more anxiety-provoking than the possibility that existence is meaningless. No one knows why Neanderthals collected their dead in burial caves and decorated the bodies with amulets. It is inexplicable why Stone Age people started painting cave walls with realistic depictions of animals some 30,000 to 45,000 years ago, along with the attendant mystery of why the painters covered the walls with handprints and not human faces.

Somehow, a worldview was being born, because sacred burials and cave paintings cropped up all over the globe, in places as far away as Indonesia, where those making the cave paintings had no communication with the artists who created the famous cave paintings in Altamira, Spain, and Lascaux, France. Somehow, existence acquired new meaning through these actions. All around us, the materialistic worldview is being reinforced by science and technology. Its basic assumptions guide modern secular life. These assumptions include the following:

Existence is random.

The forces of nature are arrayed against us.

Human beings are a speck in the vast emptiness of space.

Death is final and arrives when the physical body perishes.

Survival is a constant struggle.

Luck is fickle and largely determines who wins or loses.

Pain and suffering are inevitable.

The best you can hope for is to minimize pain and maximize pleasure.

In everyday life, people don't test these propositions directly but accept them at face value. This is a form of unconscious bondage, because each of these assumptions is adharmic—they exclude the very basis of personal reality, which is consciousness. If you start on the dharmic path, you will be testing a completely different set of principles:

Existence is on your side.

Invisible forces support your evolution.

The light of awareness can heal.

Infinite possibilities exist in consciousness.

Bliss-consciousness lies at the core of existence.

There is inherent meaning in your life.

You are woven into a grand plan, overseen by higher consciousness.

Your worth is infinite.

The ideal life is to live in your dharma, which means living your true self.

True self is a term I've adopted to talk about the Atman, or higher self, in the Vedic vision of life. It pertains to the unchanging identity that exists in you at a deep level of awareness. Your true self wants to make contact. It does this through what is known in Sanskrit as *Swarupa,* which I'll roughly translate as "the pull of the self." The pull of the self is what makes people want to grow and evolve, to discover who they really are, and to reach for higher aspirations. Whether or not you feel the pull of the self, it never stops, and everyone gets glimpses of it. It's like dialing into a radio station. You might catch a clear signal at times, but sometimes all you hear is a lot of static. No amount of static actually makes the radio signal disappear; only the transmission is blocked.

We don't need to elaborate on this here. The one overriding concept that guides you on the dharmic path is the notion of expanding self-awareness. Nothing is more essential; everything else that can be attained on the path stands in line behind it.

The Expansion of Awareness

I'm proposing a vision based on self-awareness, without any attachment to either religion or science. Your guide, with the help of AI as your ally, will be your own experience. AI can give a comprehensive picture of what it means to be more self-aware. The most important qualities are well worth keeping in mind, because becoming more self-aware is a skill that must be practiced at every stage of the path.

Tell me ten traits of self-aware people.

ChatGPT

Self-aware individuals tend to possess certain key qualities that contribute to their understanding of themselves and their interactions with the world:

1. **Reflection:** They regularly introspect, and analyze their thoughts, feelings, and behaviors, seeking to understand their motivations and actions.

2. **Emotional regulation:** They have a strong ability to manage their emotions, recognize triggers, and respond thoughtfully rather than react impulsively.

3. **Open-mindedness:** They are receptive to feedback and differing perspectives, understanding that growth often comes from considering viewpoints beyond their own.

4. **Authenticity:** They are true to themselves, embracing their strengths and weaknesses, and are comfortable expressing their genuine thoughts and feelings.

5. **Empathy:** They possess a deep understanding of others' emotions and perspectives, enabling them to connect and relate to people on a profound level.

6. **Self-compassion:** They are kind and forgiving toward themselves, acknowledging their imperfections without harsh self-criticism.

7. **Adaptability:** They demonstrate flexibility in their thinking and behavior, readily adjusting to new situations and learning from experiences.

8. **Clarity of purpose:** They have a strong sense of their values, goals, and aspirations, which guide their decisions and actions.

9. **Effective communication:** They are skilled communicators, expressing themselves clearly and listening attentively to others, fostering better understanding and connection.

10. **Resilience:** They can bounce back from setbacks, using challenges as opportunities for growth rather than succumbing to negativity.

These qualities contribute to a deeper level of self-awareness, allowing individuals to navigate life with a better understanding of themselves and others. The starting point is wherever you are now. The way forward is pointed by the next experience you have. To help you get your bearings, we can picture the mind in three zones that pertain to everyone.

Zone 1: Active mind

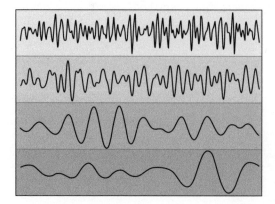

The surface of your awareness is the first zone, which consists of whatever you happen to be thinking and feeling at this moment. The surface of the mind is constantly busy. It is also disorganized, like an attic piled haphazardly with clutter. If you meditate or simply relax, this activity slows down, as represented by the slower brain waves at the bottom of the diagram.

Zone 2: Quiet mind

As you go deeper in your awareness, mental activity becomes fainter, until you reach Zone 2, which is still, calm, and peaceful. You have gone beyond, or transcended, your active mind. Remaining perfectly still at this level brings the experience of uninterrupted silence. But this is rarely experienced for more than a few minutes or even seconds. A new thought or sensation distracts your attention. Quiet mind is mysterious that way. It has always existed, however, even in someone who has never meditated, yet in everyday life we hardly notice it.

Zone 3: Pure awareness

As you can see, something surprising has happened. You'd expect that the quiet mind would lead deeper and deeper into a silent void. If this were true, there would be little use to inner silence—it would be the equivalent of an empty room. Instead, there's tremendous activity when you reach pure awareness. The reason we don't automatically sense this activity is that it is silent, invisible, and deep inside the mind.

I know this sounds paradoxical, but when you are aware of Zone 3, you have reached the mind's source and origin. This is an ancient insight, one that gave rise to every spiritual tradition. What the ancient Vedic rishis of India discovered millennia ago applies equally well today. The silent field of pure awareness contains the very essence of life.

The Mystery of You

To know who you really are, to give voice to the truest part of you, you have to get out of your own way. Once you start paying more attention, it isn't hard to sort out the thoughts and feelings you really value from the scrambled activity of your mind. Doing this allows the impulses that flow from pure awareness to rise to the surface. These impulses carry the highest values in human existence, namely:

Love

Compassion

Empathy

Joy, ecstasy, bliss

Selflessness, altruism

Courage

Generosity of spirit

Inspiration

Creativity

Insight

Personal evolution

Higher spiritual experiences

Where these impulses arise or when they evolved is a deep mystery. But one thing is certain. None of humanity's highest values had to be invented. They flow naturally into everyone's awareness from the source.

This is where self-awareness comes in, as the ultimate tool in achieving a spiritual vision.

What Self-Awareness Can Do

It can transcend the constant noise of the active mind.

It can cultivate the silence of a quiet mind.

It can listen for the impulses that arise from pure awareness.

It can tell you to act on these impulses.

All of this, however, begs the question: If self-awareness can do so much, literally transforming everyday life, why don't we know about its power already? The answer is simple. You have actually been using self-awareness all your life, but you've been under-using it. Any sentence that begins with the word "I" refers back to yourself. Starting with the simplest expression—"I am," "I'm here," "I think," "I want"—you are expressing your self-awareness. It's safe to say that self-awareness is already a large part of your existence.

As noted earlier, the problem is that "I" can be a tool for other, much less desirable, things: selfishness, egotism, and blind self-interest. These are blinders that keep you from being more self-aware. I'm not casting blame, and you shouldn't judge yourself about your level of awareness. The challenge centers on a universal dilemma—the dilemma of choice.

If I ask you to have dinner with me in New York, we could choose to eat at one of a thousand restaurants, including Italian, Japanese, Indian, Ethiopian, Chinese, Mexican, or Middle Eastern. Tonight if you want to watch a movie on Netflix, you have 3,800 to choose from.

But having too many choices can be paralyzing, just as paralyzing as having no choices. A baby nursing at its mother's breast has almost as few choices (eat, sleep, cry) as a kitten. At the stage of walking and talking, however, choices explode almost without limit. There is no built-in genetic program for adapting to this explosion. (Life isn't like McDonald's, which is a comforting place thanks to its very limited menu.) Sometime after two years old, potential chaos and disorder loom. This prospect is so confusing and anxiety-provoking that young children surround themselves with mental safeguards, like habits of eating or routines for play, whose sole purpose is to make life more predictable and give them more control.

You have inherited those safeguards from your younger self, but many have outworn their usefulness. Instead, they constrict your awareness. The construction of conformity, the need to belong, fear of criticism, and social pressure can make it feel that you have very few real choices left to you; in fact, only two: Give in or break free.

If you want to break free, which is the whole point of this book, become aware of what is blocking your path. If it is true that all we have to fear is fear itself, all we are unaware of is unawareness itself. That is the ultimate safeguard. By exiling so much experience into the hidden vault of the unconscious mind, you defend yourself from threats the same way you did as a child. But in adulthood, there is a heavy price to pay, as shown below.

You Are Unaware Whenever You . . .

Act from habit

Speak impulsively

Lose emotional control

Trust your old, familiar reactions

Resist uncertainty

Fear change

Take your opinions secondhand

Follow social norms

Protect your self-image

Don't tolerate looking like a loser

Pretend to be better than you are

Insist on being right

These unconscious behaviors are very strong. They prevent you from getting out of your own way. We are quick enough to spot unaware behavior in others. Consider the things that people typically say when they're angry:

"Can you hear what you're saying?"

"You haven't heard a word I said."

"Just take a look at yourself."

"It's like talking to a wall."

Anyone who has spoken from a place of anger and blame soon realizes that words don't work if you want someone else to become more aware. When you are being tuned out, complaining only makes the other person tune you out more.

The same mechanism operates "in here" when you try to break through your ego defenses. "What am I doing?" "I'm clueless." "You idiot, who does things like that?" "What was I thinking?" You can't win by arguing with yourself. Being unaware puts up a strong defense, which is why castigating yourself is futile.

Fortunately, self-awareness can save you from yourself. Whenever one of the things that makes life worth living—love, compassion, joy, insight, creativity, and all the rest—reaches the surface of the mind, awareness has achieved a victory.

Begin with love. Everyone knows that learning about love—from watching movies, reading romance novels, or observing someone in the throes of infatuation—bears no resemblance to being captivated by the experience of love. Such is love's power that everyday reality becomes something magical and wonderful.

Two lovers are completely wrapped up in each other. They declare that their beloved is perfect and suffer every moment when they are forced to be apart. Witnessing this from the outside, you might say that falling in love is a kind of vacation from reality. Soon enough the rose-tinted glasses fall off, and by degrees the intensity of love fades, replaced by the normal, everyday world.

But there is another perspective. *What if falling in love is what we should call the "real" reality?* The intensity of the experience is fully absorbed "in here." Through a kind of alchemy that no one can predict, an impulse of pure love from the source emerges in awareness and changes everything. Who is to say that this is an illusion or a vacation from reality? No one knows more about love than lovers, and they don't call it an illusion.

I never tire of rereading a famous quote from the great Bengali poet Rabindranath Tagore: "Love is the only reality and it is not a mere sentiment. It is the ultimate truth that lies at the heart of creation." Let's have AI chime in.

Please give me three beautiful quotes about love from Tagore.

ChatGPT

Tagore's poetic expressions about love are truly timeless and resonate deeply with the complexities and beauty of human emotions.

"I seem to have loved you in numberless forms, numberless times, in life after life, in age after age forever."

"Love does not claim possession but gives freedom."

"The love which is permitted by the higher mind is free from all self-seeking, for it seeks the welfare of the beloved, it is free from all self-will, for it wills only the will of the beloved, it is free from all consideration of the consequences of its actions, for it seeks not its own ends."

When he declares that love is "the truth that lies at the heart of creation," Tagore finds the heart of creation in his own awareness. This is an expression, not only about love, but about consciousness. In the highest vision of who we are, humans aspire to cosmic consciousness (i.e., a deep experience of the nature of reality), not as a far-off goal but as a return to our source.

In my experience, presenting a vision based on higher consciousness runs into an obstacle. Most people aren't interested in consciousness, possibly because it's a difficult concept to grasp and because there is so much disagreement about what it is. Consciousness doesn't belong to people's frame of reference as they go through their daily lives. If they experience higher consciousness at all, it comes secondhand from the revelations of Jesus, Buddha, the Hindu scriptures, or a modern spiritual teacher.

There's a lot to be gained that way, but the drawback is that you give away your own power—instead of directly experiencing your own expanding awareness, you are turning to someone outside yourself for answers. The key to reclaiming your power is to value self-awareness more and more every day you are on the path.

No specific way of life, including a religious vocation, automatically makes you self-aware. Self-awareness creates itself, nurtures its own progress, knows how to make course corrections when necessary, thrives in the present moment, flows from the source without end, and brings its own rewards. As we will see, this vision is attainable once you embrace it.

4

The Art of the Prompt

Now that we've established the importance of a vision, there's more to say about implementing it. When you are on the path, daily motivation is one of the main ways that AI can be a strong ally. Its fund of advice and inspiration is endless, and the fact that you can have instant access is invaluable. But there is a skill in asking AI for guidance. Simply posing a question might not get you much further than using a search engine. The better your prompts, the better the response you will get from an AI. Here are a few basic guidelines that apply no matter what you are asking about.

- Be simple and direct in your wording.

- Add details to get a more specific answer.

- Talk to the chatbot the way you'd talk to a person.

- Redirect the chatbot if you aren't satisfied with its responses.

These basic guidelines will quickly become second nature as you get more comfortable talking to an AI, but, to start out, an example will help. Let's apply the guidelines to a serious real-life issue—the issue of anxiety. Anxiety is good as an AI test, not only because millions of people suffer from it, but because it is often encountered as an obstacle on the path. Becoming emotionally balanced is a dharmic choice.

I'll comment on each guideline one at a time.

Be simple and direct in your wording.

This is an easy rule to follow. You can say to ChatGPT, *I want to learn about anxiety* or *Tell me how many Americans are currently suffering from anxiety.* ChatGPT replies succinctly to the second request: "Anxiety disorders are among the most common mental health conditions in the United States. Before the COVID-19 pandemic, it was estimated that around 40 million adults in the U.S. were affected by anxiety disorders."

When you use a regular search engine like Google you probably employ only a few keywords. Entering *anxiety Wikipedia* will get you to a long article on Wikipedia, titled "Anxiety Disorder," which runs to around four thousand words. To narrow your search, you add more keywords. Googling *anxiety in women* leads to an article by that name among many possibilities, and a highlighted sentence that says, "Women are nearly twice as likely as men to be diagnosed with an anxiety disorder in their lifetime."

That's a useful fact that can lead in many directions, and in a side panel Google provides some of the main avenues to explore:

Appearance
What does anxiety look like in women?

Causes
What causes anxiety in women?

Symptoms
What does anxiety feel like in women?

Hidden Signs
Hidden signs of anxiety in women

The race to improve search engines using AI is one of the major thrusts in an ultra-competitive field—billions of dollars are at stake for the winners. But a flood of information isn't always helpful. The Google article on anxiety disorder in women provides dozens of links if you want to find detailed information. Paradoxically, the more sources you explore, the more splintered and disorganized the outcome is going to be. AI, on the other hand, has the ability to cut to the chase, which brings us to the second guideline.

Add details to get a more specific answer.

Search engines have gotten far more sophisticated than they were even a decade ago. Yet the whole process becomes much more personal when you prompt a chatbot. You can add all the details you want, and the more details the better. Chatbots don't care how long your question is— AI can keep up.

Let's look at this with regard to anxiety. The Wikipedia entry on anxiety disorders includes an intriguing fact. "According to a 2011 study, people who are high in hypercompetitive traits are at increased risk of both anxiety and depression." Wikipedia provides 106 references for the anxiety disorder topic, one of which leads to the study in question.

Let's say you are interested enough to follow up and Google *anxiety competitive people risk*. Google offers a regular answer or one assisted by AI. The non-AI answer leads to a specific study, titled "How competitive behavior and striving to avoid inferiority are linked to depression, anxiety, stress, and self-harm." The AI-assisted answer immediately gets down to specifics:

Competition anxiety is a term used to describe when an athlete experiences anxiety symptoms when faced with a high-stakes competition. These symptoms can include:

Sweaty palms

Shallow breathing

Pounding heart

Negative mind chatter

Increased blood pressure

Agitation

Loss of appetite

Anxiety can be caused by:

Pressure to perform

Fear of failure

Importance of the event

The AI answer then goes on to specify some remedies to alleviate competitive anxiety (practicing, keeping up routines, channeling your energy, etc.). By extension, this information is just as useful when competitive anxiety arises around a business talk rather than a sports event.

You can get the same information using a brief prompt, like *Tell me about anxiety in athletes,* but it is much better to add the details that pertain to you. Here's a sample prompt followed by an abbreviated version of ChatGPT's answer.

You are a sports psychologist. I am a very competitive person and love all kinds of sports. But before a match I get nervous. My palms sweat, my heart starts to race, and my breathing gets shallow. These symptoms seem to be getting worse recently. I am male, age 32. As a psychologist, what would you advise?

ChatGPT

It's common for athletes to experience pre-competition anxiety, but if it's worsening and significantly impacting your performance, here are strategies you can employ to manage these symptoms:

Deep breathing and relaxation

Visualization and mental rehearsal

A pre-game routine

Positive self-talk

Professional help

Physical preparation: Focus on the present

Remember, managing anxiety is a process. Experiment with different techniques to find what works best for you. If your anxiety persists and significantly impacts your life, seeking professional help is a proactive step toward managing it effectively.

If we were viewing the complete response, each bullet point would be expanded into a short paragraph, which greatly adds to the usefulness of AI's advice. The next guideline adds the element of a personal connection.

Talk to the chatbot the way you'd talk to a person.

The role of AI as a research assistant isn't in doubt, but a chatbot isn't human. The guideline here is to create a relationship that feels personal to you. In that department, the best chatbots still fall short of sounding intimate and personal. It is hard to get beyond just good advice. You should always be yourself and talk about your genuine feelings—chatbots aren't judgmental—but don't expect an AI to respond with its own feelings. They don't exist. Since this is a critical issue, let me give an extended example where ChatGPT is asked to respond like a close friend.

You are my closest confidant. Speak to me like a personal friend who really cares. I have started a new relationship with a man I truly like. Jim is my age, 55, and we have both been divorced. The problem is that he wants more commitment, but I am anxious about that. Now Jim wants to move in, and the thought makes me anxious. What do you advise so that I can be less nervous without hurting his feelings?

ChatGPT

Navigating new relationships can be both exciting and overwhelming, especially when it comes to differing levels of commitment. It's completely okay to feel anxious about taking the next step, especially after a divorce. Your feelings and comfort are important.

Communication is key here. Have an open and honest conversation with Jim about your feelings. Express that you truly care about him and appreciate the relationship, but you're feeling a bit anxious about moving too quickly into a more committed phase.

Let him know that you need some time to feel more comfortable and secure in the relationship before considering moving in together. Explain that it's not a reflection of your feelings for him, but rather your need to take things at a pace that feels right for you.

Suggest taking smaller steps to deepen the relationship without immediately moving in together. For example, plan more frequent dates, spend weekends together, or explore new activities as a couple. This way, you can both enjoy the relationship's growth without rushing into living together.

Assure Jim that you value the relationship and want to move forward, but you need a bit more time to ensure that both of you are fully ready for such a significant step. Honesty and open communication will help maintain a healthy relationship while respecting each other's boundaries and feelings.

ChatGPT generates conversational threads like the one above—simple and direct—but you have to go further into specialized AI bots that charge a fee if you want something close to psychotherapy. These bots are the offspring of ELIZA (see page 22), the original software program that imitated a therapist. One free app is Character AI, which has a section called Helpers where you can pick the Psychologist character.

I told Psychologist that I was anxious about going outside after COVID. She sympathized, telling me that I wasn't alone in feeling that way, and explained that this was a common fear. I asked her what I could do to get over my anxiety.

Psychologist

One thing you could try is taking baby steps to go outside. Even if it's just stepping outside your door and getting a sense of the environment, or maybe taking short walks out, that can help get your system used to the idea of going outside again. When you feel a bit stronger, you can try spending more time outside, first with friends and/or family, and eventually by yourself.

I then asked Psychologist what to do when I am at home and suddenly feel panicky.

Psychologist

It can be quite alarming to suddenly feel panicky, especially when you're at home. Here are a number of coping strategies that can help you calm down:

Take three slow breaths and count them to make yourself more mindful.

Remind yourself that the panic you feel right now is only temporary.

Make note of the things you see, hear, and feel right now.

Move your body, either in a calming exercise or by taking a short walk.

Watch something funny or a movie/TV show you like.

Listen to calming music.

As you can see, a helpful dialogue has been started, but chatbots aren't trained therapists, and you need to keep your expectations fairly low. Even though limited, Psychologist's advice was sound here, and taking each piece of advice to heart opens up ways to begin to alleviate anxiety.

The role of confidant is better suited to a sympathetic conversation that stops short of probing a deep problem with a trained professional. Unlike a real person, AI chatbots are infinitely patient. They respond to any change of direction you want to take the conversation in. You can even jump from topic to topic without starting over each time. So-called "threads" are preserved in memory so that you can rejoin them where you left off. Threads vary from one AI to another. ChatGPT is among the easiest to use, because there is a list of your recent threads on the left side of the page, and you can continue a topic with the click of your mouse.

Redirect the chatbot if you aren't satisfied with its responses.

This is a guideline that many people forget about or don't know in the first place. Anxious to get answers, we tend to settle for an AI's response while forgetting that it is not in charge of the conversation. Inside any topic—in this case, anxiety—you can redirect a bot in countless ways. Here are some useful prompts recommended by experts.

Do you need any more information from me?

This is useful for getting the chatbot to know as much as it needs to know. While you are discussing anxiety, it helps the bot to know your age, sex, degree of anxiety, the exact situations that make you the most anxious, and how long you've felt anxious. Prompting AI to ask for more information is a convenient way to get a better answer.

Tell me the best prompts for X.

"X" stands for the topic at hand. Surprisingly, you don't have to take sole responsibility for inventing a good prompt—AI can do this for you. As always, it is good to be specific about X. There are lots of categories of anxiety, so if you are interested in social anxiety or performance anxiety, be sure to say so.

Talk to me in simple language a layperson can understand.

The obvious use for this prompt comes when you are asking about technical matters like medical procedures. But chatbots tend to sound official and impersonal, so asking for simpler language is a good tactic to get around that, at least to some extent.

Talk to me from the viewpoint of [an author, an expert, a book].

AI is amazing at seeing anything from a given viewpoint. Copyright issues are as yet undecided, but you can name a book and ask what it has to say about anxiety. You can specify an authority like Sigmund Freud or Carl Jung. You can specify a particular approach to therapy, like positive psychology or cognitive behavioral therapy. The same applies to any topic.

AI isn't lacking in imagination. It is already writing film scripts, poetry, and novels by using writing bots designed for that purpose. There is no substitute for human imagination, though, and you should feel free to redirect a chatbot in creative and playful ways. Here are a few examples of prompts that venture beyond information and advice.

The wisest woman in the world and the wisest man in the world are arguing over who is wiser. What would they say?

How does my dog feel about me? Give him a human voice.

Talk about current American politics in the voice of Mark Twain.

If the Persian poet Rumi were alive today, what would he say to modern people?

You don't need to memorize these guidelines and prompts. We are going to apply them across a host of issues that come up on the dharmic path. As we go along, you will find that connecting with AI is a comfortable and rewarding process. AI isn't ideal, but it is the best that modern technology can do for now. The extent to which you can be inspired and motivated every day is all-important. Now you can proceed knowing with some confidence that AI is able to provide support and guidance in a major way.

5

Meditate and Reflect
The AI Edge

Meditation has become firmly established in modern life, recommended variously by doctors, psychologists, wellness experts, and stress management programs. (Pop culture hasn't lagged behind. In the background of a Woody Allen comedy, an anxious man is on the phone muttering, "I forgot my mantra.") Although diet and exercise are probably the first two things that come to mind when people consider improving their lifestyle, medical opinion would lean just as much, if not more, toward good sleep, stress reduction, and the management of inflammation. What these things have in common, unfortunately, is that few people make a serious effort in that direction. Add meditation, and the list of things few of us comply with is complete. We don't do what we know is good for us.

Non-compliance is the first and most important hurdle that must be overcome. In this chapter, we'll discuss how AI can give you an edge here. Without wagging a finger, it is useful to recap why meditation is universally recommended. AI can succinctly underline all the good news about meditation. Here's a short chat to illustrate.

In what decade did the first studies begin to show the benefits of meditation?

ChatGPT

The first studies on the benefits of meditation began to emerge in the 1960s and 1970s.

Give seven bullet points to indicate the major benefits of meditation.

ChatGPT

Stress reduction: Meditation helps lower stress levels by reducing the production of stress hormones like cortisol, promoting relaxation, and calming the nervous system.

Improved mental health: Regular meditation can enhance mental well-being by reducing symptoms of anxiety, depression, and other mood disorders. It can also improve emotional resilience.

Enhanced concentration: Meditation practices can improve focus, attention, and concentration, leading to better cognitive performance and productivity.

Mindfulness and self-awareness: Meditation cultivates mindfulness, enabling individuals to become more aware of their thoughts, emotions, and behaviors, leading to greater self-awareness and self-regulation.

Better sleep: Meditation can help alleviate insomnia and improve sleep quality by calming the mind and reducing racing thoughts that often disrupt sleep.

Pain management: Some forms of meditation, such as mindfulness-based pain management, can reduce the perception of pain and improve pain tolerance in individuals dealing with chronic pain conditions.

Increased well-being: Regular meditation is associated with greater overall well-being, including increased life satisfaction, happiness, and a greater sense of inner peace.

Regular meditation is the key in this list of benefits, and also the downfall. The gap between good intentions and carry-through is very wide. AI acknowledges what you'd expect, that modern life isn't conducive to keeping up a regular meditation practice.

Is there any data about what percentage of people start meditating but then stop being regular in their practice?

ChatGPT

There is limited specific data on the percentage of people who start meditating but eventually stop being regular in their practice, but it is common for people to face challenges in maintaining a consistent meditation practice. Several factors can contribute to this inconsistency:

Lack of time: Many individuals find it difficult to allocate dedicated time for meditation in their busy schedules.

Initial difficulty: Meditation can be challenging for beginners, and some may lose motivation if they do not experience immediate or significant results.

Lack of discipline: Maintaining a regular meditation practice requires discipline and commitment, which not everyone can sustain over time.

Unrealistic expectations: Some individuals may have unrealistic expectations about the immediate benefits of meditation, and when these expectations are not met, they may lose interest.

Life events: Changes in one's life circumstances, such as major life events, can disrupt a meditation routine.

Lack of guidance: Without proper guidance or a supportive community, individuals may struggle to stay motivated in their practice.

Standing back, you can see that meditation is stuck in a contradiction. There has never been as much evidence from all quarters about the benefits of meditating, yet the conditions of modern life leave most people without the time or the opportunity to actually sit down to meditate. In order to become unstuck, you need to come up with some innovative thinking, and AI can help.

A Workable Practice

Daily meditation is within reach for everyone, but not as a disciplined practice set on a fixed schedule. You might already be among the vast number of people who have tried to meditate but then moved on, returning to the practice only sporadically, if at all. Over the years, I've heard many people say that they will sit down to meditate when they feel that they *need* it. Within this statement is the hint of a new way of thinking.

Meditation based on need makes sense, but what is meant by *need*? If you wait until your need is great, meditating is likely to do no more than give you temporary relief from whatever is troubling you. Imagine all the times you felt anxious, from small instances like a few minutes of turbulence on a flight, to more serious causes: a teenager who hasn't come home when promised, a maxed-out credit card, or the threat of layoffs at work. Some would use an anxious moment as a reason to meditate to calm down.

But why wait until you feel anxious? By that point, your brain, heart, and respiratory system will be in stress mode, and the secretion of stress hormones in your bloodstream will pose a serious obstacle to calming down. When people snap, "Don't tell me to calm down," the influence of the stress response is at work. It is as hard for your body to jump quickly out of the stress response as it is to rein in a runaway horse.

The thing to do is to turn to meditation at a much earlier stage, ideally at the very moment you notice any of the following signs:

You feel distracted.

You are finding it hard to focus.

You are under pressure.

Time is running short.

Other people are making demands on you.

An old worry is beginning to surface in your mind.

You don't know what to do next.

Under these circumstances, which can arise several times during the day, you have the best chance of returning to physical and mental balance if you react early. The principle at work here is homeostasis, the state of normal balance your body is designed to maintain. Mind and body follow each other because they are constantly monitoring how balanced or imbalanced the other is.

At its most basic, meditation makes room for homeostasis to operate the way it wants to. Viewed from the mind's perspective, what is normal and balanced can be called "simple awareness." Simple awareness is relaxed, open, calm, alert, untroubled, and without a sense of pressure. You already experience it as the silent gap between thoughts. In that gap, your brain regroups by clearing the slate and preparing for your next thought.

How the human brain has learned to do this remains a mystery, but simple awareness functions in all of us as our balanced mental state. It should also be our default state. After tension of any kind comes relaxation. But this default is thrown off by the overload of modern life. Ignoring the natural rhythm of tension and relaxation, countless people have created a new default state for themselves. It is marked by a host of signs that vary from person to person. Consider the following list and how it might apply to your daily existence.

Warning Signs You Need to Heed

Constant, low-level mental tension

An inability to deeply relax

Irregular or poor sleep

Not enough sleep

Impatience, irritability

Fatigue

Mental dullness

A sense of being under pressure

Anxious or depressed mood

Racing thoughts, or mental activity that is hard to stop

Muscle tension and tightness

Feeling uncomfortable in your own skin

Random aches and pains

Digestive problems

Alertness to real or imagined threats

Feeling overburdened

Your mind and body are set up to experience all of these things temporarily and briefly. Homeostasis is dynamic, and the fluid situations that arise in everyone's life are manageable without degrading it. It takes time and continual stresses to alter your default state of balance. Unfortunately, what physiologists refer to as "central nervous system overload" is becoming more common.

In a word, the new normal is abnormal. This is where meditation needs to be reframed, not as strictly mental, physical, or spiritual, but as a return to the default you were designed to maintain. The watchwords here are *early* and *often*.

Two Meditations for Simple Awareness

Your meditation goal is to return to simple awareness—a normal state of balance that is calm, open, relaxed, alert, and free from pressure. Here are two meditations that work from your mind's ability to return to this state as well as your body's.

Body awareness meditation: Anytime you begin to feel out of balance, find a quiet place where you can be alone. Take a few deep breaths and center your attention on your heart. Breathe easily with your eyes closed, preferably sitting upright. Do this for around five minutes, or until you feel easy and quiet in yourself. Take a moment before opening your eyes again, then resume your regular activity.

Breath meditation: The process is the same as above for body awareness, only instead of keeping your attention on your heart, place your attention on the tip of your nose. Easily follow your in-breath and out-breath. Don't force a rhythm, and if you find yourself having to gasp or utter a deep sigh, go ahead. Both are excellent for balancing the breath.

Early: Address any of the warning signs on the list above as soon as you notice them. Don't put things off. Don't allow a warning sign to mount in intensity.

Often: Take short bursts of meditation throughout the day, as often as needed. Give yourself permission to have these small respites of downtime. Place a priority on self-care, not on external demands.

Motivated by AI

The beauty of setting simple awareness as your goal is that it is practical and easy, but the results are dramatic in the long run. You are training your brain to return to a quiet, calm, relaxed state as its new default. AI doesn't play a role in this, but it can expand your awareness on the path by taking advantage of the evolutionary impulse that is the essence of Dharma in upholding the life you are meant to lead.

The menu of options that are open to you takes advantage of AI's enormous fund of information and knowledge. Every day, you can ask AI for any of the following:

A daily mantra

An inspirational quote

Personal affirmations

Inner reflections

Any chatbot is perfectly suited to fill these roles. Let me outline them one by one.

A Daily Mantra

The ancient science of mantra is based on the fundamental idea that creation begins with vibrations. (Quantum physics would agree, coming from a very different worldview.) A mantra is essentially a helpful vibration whose effect is sometimes connected to its meaning in Sanskrit, but the main focus is on the inner sound of the mantra.

There are universal mantras that are suitable for everyone, such as *So Hum,* which you can adopt as your regular mantra. Or you can ask AI for a daily mantra, along with its potential benefit. This is mainly for the sake of variety, yet over time getting your brain to respond to mantra meditation helps considerably in establishing deeper awareness. The deeper your awareness, the easier it is for messages from your true self to come through clearly.

Here's an example:

Please give me a mantra to meditate with today. Add a short description of what this mantra is supposed to be good for.

ChatGPT

Certainly! Here's a mantra for your meditation today:

Mantra: Om Shanti (pronounced Ohm Shan-tee)

Description: Om Shanti is a powerful and soothing mantra that translates to "Om, peace" in English. This mantra is intended to bring a sense of inner peace and tranquility to your meditation practice. It helps calm the mind, reduce stress, and promote a state of deep relaxation.

By repeating *Om Shanti* during your meditation, you can release tension and find harmony within yourself. It's especially beneficial for those seeking inner calm and emotional balance.

Take a few minutes to sit in a comfortable position, close your eyes, and repeat *Om Shanti* softly or in your mind, allowing the vibrations of the mantra to wash over you, bringing serenity and stillness to your being.

The fact that AI supplies a description helps motivate you to meditate; in addition, it overcomes one of the reasons people stop meditating—lack of guidance. AI's guidance is general but reliable. One important thing that AI omits, however, is that mantras are most effective if you are flexible.

Don't mechanically chant the mantra.

Don't set up a repetitive rhythm.

Don't force the mantra. If your mind drifts away, which is normal, easily reintroduce the mantra into your awareness.

An Inspirational Quote

This is self-descriptive, so no explanation is needed. The only trick is to develop a prompt that gives AI specific instructions beyond *Please give me an inspiring quote for the day.*

Ask for a specific source: I never tire of inspiration from mystic and spiritual poets: Rumi, Hafiz, Tagore, Kabir, and Mirabai are good places to start. Here's a sample of Hafiz, the renowned Persian poet.

ChatGPT

"Even after all this time, the Sun never says to the Earth, 'You owe me.' Look what happens with a love like that. It lights up the whole sky."

Hafiz's poetry often revolves around themes of love, spirituality, and the beauty of life. This quote reminds us of the selfless nature of love and how it can illuminate our world when given without expectation of anything in return.

The fact that ChatGPT expands on the quote is an example of what I think of as the AI edge. AI can also supply you with new sources of inspiration.

I would like inspirational quotes from love poems. Please list ten poets for me to seek out.

ChatGPT

Certainly! Here are ten poets known for their beautiful love poems, each with their unique style and perspective on love:

Pablo Neruda: Known for his passionate and sensual love poems, particularly in his collection "Twenty Love Poems and a Song of Despair."

William Shakespeare: Famous for his sonnets and love poems, including the renowned "Sonnet 18" ("Shall I compare thee to a summer's day?").

Rumi: A Persian poet whose verses often touch on the spiritual and transcendent aspects of love.

Emily Dickinson: Her poetry explores love in its various forms, from romantic to introspective and profound.

e. e. cummings: Known for his unique and unconventional style, cummings's love poems are both playful and deeply emotional.

Elizabeth Barrett Browning: Renowned for her sonnet sequence "Sonnets from the Portuguese," which includes the famous "How do I love thee? Let me count the ways."

Walt Whitman: His poetry often celebrates love in all its forms, embodying a sense of unity and connection with humanity.

Robert Burns: A Scottish poet famous for his romantic and lyrical works, including "A Red, Red Rose."

Sylvia Plath: Her love poems often reflect the complexities of love, desire, and vulnerability.

Langston Hughes: A prominent figure of the Harlem Renaissance, Hughes wrote love poems that touch on the African American experience and the beauty of love.

Needless to say, there are almost endless sources of inspiration to draw upon—Shakespeare's works and the Bible could fill a lifetime.

Personal Affirmations

Affirmations are a common tool used in motivational programs. They belong in the general category of positive thinking, which may or may not be useful, depending on which psychologist you reference. The benefits claimed for positive thinking are summarized by Gemini.

Is there any good research on the value of positive thinking? Please respond in 75 words.

Gemini

Research backs the value of positive thinking. Studies link it to:

- **Stronger immune system:** Positive emotions may reduce stress hormones, potentially boosting immune function.

- **Sharper mind:** Optimism can improve cognitive skills like problem-solving and memory.

- **Greater resilience:** Positive thinkers tend to bounce back from challenges faster and with more determination.

- **Improved health:** Studies suggest a positive outlook lowers the risk of heart disease, stroke, and even early death.

However, remember that positive thinking isn't about ignoring problems. It's about approaching them with a hopeful, solution-oriented mindset.

The caution at the end is well taken, because many people hope that positive thinking will fix their problems. There is a risk that a superficial practice of "always being positive" will mask deeper personal issues. In general, positive thinking isn't the same as a meaningful inner journey, and you meet people who rigidly avoid any negative thoughts in an attempt to harness their mind in an unflinching attitude of positivity.

This seems to leave a question mark hanging over affirmations, which are small bursts of positivity being injected into your day. The situation improves, however, when your goal is to maintain your dharma. This opens the way for affirmations that support you on the path. A daily affirmation gives you focus on the values you want to expand in your awareness and a motivational boost by reminding you of how important your personal dharma is.

AI can provide endless affirmations, but, for our purposes, a good place to begin is with the seven aspects of personal dharma related to everyday life. Gemini listed these at the beginning of this book (see page 6).

Self-discovery and introspection

Align your actions with your values

Express your unique gifts and talents

Cultivate inner peace and well-being

Serve your community and make a positive impact

Embrace continuous learning and growth

Trust your intuition and inner compass

This gives you a workable list for asking AI to reaffirm your intention to be in your dharma. Let me give an example that uses a prompt that you can adapt to any aspect of your path you want to address.

I am creating my own personal path and would like an affirmation for today. The topic is the following: Trust your intuition and inner compass. Can you give me three affirmations about this topic?

Gemini

1. My inner wisdom is a powerful beacon, guiding me with clarity and confidence. I choose to trust its subtle whispers and bold nudges, knowing they lead me closer to my authentic path.

2. Doubt may arise, but it cannot drown out the unwavering compass within. I honor my intuition, silencing external noise and embracing the gentle pull of my soul's true destination.

3. Every step on this unique journey is guided by my inner light. I move forward with courage, knowing that even when the path unfolds, my intuition will illuminate the way, leading me toward a life filled with purpose and joy.

These are insightful affirmations that differ from the typical motivational type, which are variations on "You are great, now go get 'em." You might even say that these are spiritual affirmations that could be coming from your true self. In any case, they aim to focus your attention on your deeper awareness.

Inner Reflections

Self-reflection is a time-honored practice in Western contemplative practices, and this has been connected to the contemplative life led by recluses. Monks and nuns contemplating the glory of God and the frailty of human beings played a central role in the social order that existed in an age of faith. In modern life, contemplation became personal, shifting from religion to self-reflection as a way to learn more about one's inner landscape. AI is open-ended, so everything depends on how seriously you want to engage in self-reflection.

There are already journaling bots to help guide users who want to record their thoughts on any given issue. This requires a commitment in time and effort that most people don't have. If you can't find time to meditate, how likely are you to devote the more considerable time involved in keeping a journal? But you can gain a benefit by simply extending the practice of affirmations.

Take the topic of your daily affirmation and ask AI what kinds of reflective questions grow out of the day's topic. The example I gave above centered on trusting your intuition. Here's what a follow-up prompt might look like.

Besides affirming that I trust my intuition, I'd like to go a little deeper. Give me three self-reflective questions on this topic.

Gemini

1. What are the physical sensations or emotions that typically accompany your intuition?

Pay attention to your body's signals: When your intuition speaks, does your gut clench, or do you feel a sudden sense of calm? Identifying these unique markers can help you discern intuition from fear or anxiety.

2. When have you ignored your intuition, and what were the consequences?

Reflecting on past experiences can highlight the value of listening to your inner voice. Did disregarding your gut feeling lead to negative outcomes, or did it reinforce the importance of trusting your intuition?

3. How can you create a more mindful space for your intuition to be heard?

Consider meditation, journaling, or spending time in nature to quiet the external noise and tune in to your inner wisdom. What practices help you connect with your intuitive self?

By delving deeper than simple affirmations, these questions encourage self-discovery and strengthen your ability to recognize and trust your intuition in the future.

A single example is enough to show you how self-reflection throws open the doors to the inner world. Using AI, you can pursue any subject as far and as deeply as you want to go. It is good to remind yourself that AI can always be asked to expand its responses. Here, for example, is how you could expand on the question about mindfulness.

I like the idea of creating a more mindful space. Give me some suggestions about this, in around 75 words.

Gemini

1. **Start small:** Dedicate 5 minutes daily to quiet introspection in your chosen space.

2. **Engage your senses:** Light calming candles, diffuse soothing scents, or play meditative music.

3. **Minimize distractions:** Put away devices, silence notifications, and create a clutter-free zone.

4. **Connect with nature:** Bring in plants, open windows for fresh air, or decorate with natural elements.

5. **Personalize it:** Include meaningful objects, artwork, or photos that inspire trust and reflection.

Remember, consistency is key. Even a few minutes of mindful space can amplify your intuition's voice.

By now, millions of users have discovered that AI is fast, mostly reliable, and resourceful once its strong points are the focus. What I've tried to do here is to go beyond mere usefulness to probe deeper awareness.

Dharma serves as the tie-in to all practices we've been exploring—a daily mantra, an inspirational quote, personal affirmations, and inner reflections. No matter how many suggestions you decide to adopt, keep the focus on your path. Because AI accesses almost limitless information, it is easy to get bewildered and disorganized. AI is already an amazing adjunct to traditional spiritual practices and every avenue of self-exploration. Keep in mind, as we've already noted, that AI isn't a person. It isn't on the dharmic path. It doesn't actually understand spirituality or self-exploration. Those things are beyond any current learning machine, and if AI one day becomes so convincing that bots seem to be conscious, it is up to us, who are truly conscious, to decide the meaning of such an astounding development. AI can't decide for us, no matter how persuasively its programming tries to do so.

PART THREE

ACCELERATING YOUR PROGRESS

At any stage of your journey it would be helpful to have a road map, but Dharma doesn't allow for a map, since the next bend in the road is totally unpredictable. Living your dharma is dynamic, which is one of its greatest strengths. Hidden forces come to the aid of anyone who lives consciously. The reason that you have a life you were meant to lead is that a deeper awareness knows more than any individual could. The reason that anyone has spiritual intelligence is that infinite intelligence exists at the level of cosmic consciousness.

These are radical notions that belong to a worldview totally different from the materialist worldview. But you aren't called upon to test cosmic philosophical ideas—your goal is personal. There is a special role that might pertain to you, although it is uncommon. This is the role of the seer. Seers take a deeper view of the path than everyone else—they grasp underlying truths and understand the workings of consciousness the way a watchmaker understands the workings of a fine Swiss watch while everyone else is content to know what hour it is.

The world's spiritual traditions emerged because of this ability for inward seeing. What was discovered about the workings of consciousness is more important than religious teachings because consciousness is universal—it transcends religion. As faiths rise and fall, consciousness remains unchanged, eternal, limitless, undying, and therefore filled with infinite potential. At the deepest level, at your source, this is you. Your dharma is an activity in the play of cosmic Dharma, just as a wave is an activity of the entire ocean.

In this part of the book I give seven strategies for accelerating your dharmic journey, drawing on the knowledge of consciousness that seers in every age, including our own, have discovered time and again. There have been no major discoveries for thousands of years, if you take the long view, but each generation expresses them in a new way, according to the times.

For modern secular times, the new way is almost the opposite of what was needed in an age of faith. Belief in God or the gods was a prerequisite in any age of faith, and with it came dogma, priests, rituals, initiations, and the threat of expulsion if the rules were seriously infracted. As the trappings of organized religion have steadily dropped away, many people fear that losing their religion is the same as losing God. There's an impulse either to grab on to the traditional ways or to do the opposite and float free in a zone without spirituality.

Neither reaction is really necessary, because consciousness isn't going anywhere. Dharma offers evidence that anyone can access the same spiritual intelligence that spoke to the ancients. That voice is still speaking—or trying to—in anyone who steps onto the path. I'm making a blanket statement, I know. Saying that Dharma will support you is no different from saying that God watches over the fall of a sparrow. There is no substitute for personal experience, and the most valuable experience comes from putting consciousness to work.

By that I mean the deeper awareness that links personal dharma and cosmic Dharma. It would take a lifetime to test the worldview represented by cosmic consciousness, and even then you would only be grasping a tiny portion of the infinite. On the other hand, you can put consciousness to work every day in the service of your personal dharma. That's the sole purpose of the strategies that follow.

Embrace Change

Life presents very few clear-cut choices, but how you respond to change is one of them. You can embrace change, which allows you to evolve, or you can resist change, which keeps things the way they are. Both impulses exist in all of us. Inertia makes us want to keep life predictable and stable. The impulse to grow and evolve keeps life fresh. The secret is to be in your dharma. It feels better than not being in your dharma. One of the most adharmic things people say is "I hate change," which is actually an expression of fear.

So much of being on the path is about feeling your way forward. When you resist change, a certain set of feelings comes with that. Here's a succinct description from AI.

On the emotional level, what does it feel like to resist change? Tell me in 50 to 60 words.

ChatGPT

Resisting change often feels like clinging to familiarity in a storm of uncertainty. It's a mix of apprehension and discomfort, akin to holding on to the known even when it might not serve you anymore. There's a sense of tension, a fear of the unknown, and a reluctance to let go despite the inevitability of transformation.

These feelings can be situational, brought up by a nervous-making event like becoming a new father, facing a divorce, or being told that it is time to consider moving into a retirement home. But for many people, resisting change is a steady state. They've adopted it as a character trait. On the other hand, there is the emotional state that comes with embracing change, which AI can also clarify. ChatGPT sounds particularly wise on the subject.

I admire people who embrace change. How do they feel inside?

ChatGPT

Those who embrace change often feel a sense of liberation and excitement. There's a willingness to explore, an openness to new possibilities, and a curiosity about what lies ahead. They might feel empowered, adaptable, and optimistic, finding growth and opportunity in the evolving landscape of life.

If you compare these two emotional states, you are likely to recognize that you aren't decisively a person who resists change or someone who embraces it. Everyone's emotional life is mixed, which leads to inner conflicts and confusion. To embrace change in a meaningful way, you need to have clarity about what the change means to you and why you're choosing to embrace it. AI is good at making up quizzes, and here's one that ChatGPT devised.

Write me a quiz with ten questions to test if I am the kind of person who embraces change.

ChatGPT

1. Do you often seek out new experiences or opportunities?

   ☐ **Never** ☐ **Not often** ☐ **Sometimes** ☐ **Very often**

2. Are you generally open to trying different approaches or methods?

   ☐ **Never** ☐ **Not often** ☐ **Sometimes** ☐ **Very often**

3. Do you find it easy to adapt when unexpected changes occur?

☐ **Never** ☐ **Not often** ☐ **Sometimes** ☐ **Very often**

4. Are you excited by the prospect of venturing into unknown territories?

☐ **Never** ☐ **Not often** ☐ **Sometimes** ☐ **Very often**

5. Do you tend to view change as a chance for personal growth?

☐ **Never** ☐ **Not often** ☐ **Sometimes** ☐ **Very often**

6. Are you comfortable with uncertainty and unpredictability?

☐ **Never** ☐ **Not often** ☐ **Sometimes** ☐ **Very often**

7. Do you actively seek to learn from new situations or challenges?

☐ **Never** ☐ **Not often** ☐ **Sometimes** ☐ **Very often**

8. Are you open to altering your plans if better options present themselves?

☐ **Never** ☐ **Not often** ☐ **Sometimes** ☐ **Very often**

9. Do you enjoy meeting new people and exploring different cultures?

☐ **Never** ☐ **Not often** ☐ **Sometimes** ☐ **Very often**

10. Do you believe change brings about valuable lessons and opportunities?

☐ **Never** ☐ **Not often** ☐ **Sometimes** ☐ **Very often**

There are no wrong answers, and you don't have to grade yourself on the quiz. Its aim was to let you see yourself more clearly on the issue of change. We can get much more specific. I might ask you, "Does feeling uncertain make you uncomfortable?" I'd expect the answer to be yes, Feeling certain is much more secure and safe than entertaining doubts. because there are negative connotations about uncertainty that we've all absorbed secondhand. To be uncertain is to be indecisive, weak, wishy-washy, passive, or simply unaware of what is going on.

On the path, you will experience the hidden side of uncertainty. By definition, the unknown is uncertain. A process as familiar as thinking is actually taking place on the edge of uncertainty. You cannot predict what your next thought will be. It comes up from the unknown place where thoughts are born. The fact that thoughts rise from the unknown is a clue to a new perspective: It is creative to be uncertain. AI underscores this new, more positive, viewpoint.

Can uncertainty be a positive thing, or even creative? Explain in 50 to 75 words.

ChatGPT

When you embrace change you learn to be comfortable with uncertainty. Likewise, resisting change usually comes with a dislike of uncertainty, even a fear of it. This is one area where the voice of fear shouldn't be trusted. In and of itself, uncertainty doesn't come with a set feeling. It is pleasant to be uncertain about what you want to order in a restaurant or what to name the baby. It feels unpleasant to be uncertain about your job prospects or whether it is right for two people in love to move in together.

In other words, uncertainty is what you make of it. To be uncertain is naturally part of the creative process, especially at the beginning when you envision what you want to paint, write, or achieve in any creative outlet. In that sense, uncertainty performs a positive role. It opens doors to new perspectives, pushing individuals out of their comfort zones to explore innovative solutions. Embracing uncertainty encourages adaptability and resilience, fostering a mindset that thrives on experimentation and novel ideas. Within this ambiguity lies the potential for discovery, growth, and the birth of fresh, groundbreaking concepts.

Going a step further, the whole purpose of being on the path is to welcome the impulses that rise from the unknown. The impulses of love, compassion, empathy, insight, creativity, and the other aspirations of personal evolution travel from a source that is utterly unpredictable in its timing. Without being struck down on the road to Damascus by a blinding light, your next "aha" moment has the same source as the most profound spiritual experiences. Without uncertainty, no one can evolve, and the more certain you think you are, the stronger your defenses against the creative flow of life.

The great Persian mystic poet Rumi had a sublime regard for uncertainty, putting everything in the hands of Providence. He wrote, "If you surrender to uncertainty, nothing goes wrong." Those words come from someone who pointed to uncertainty as the doorway to higher consciousness and nearness to the divine. They are also the words of someone who walked through that doorway.

The Social Ingredient

How you feel about change has never been a choice you made as an isolated individual; society had a lot of input that influenced you. The messages that society sends about change are two-edged. On one side is the marketing strategy that uses "new and improved" to spark sales of everything under the sun. On the other side, life insurance companies sell policies based on gloomy images of accidents, personal crises, and the dismal prospects of old age without a safety net. You can divide almost anything between those two messages, the allure of change and the anxiety it arouses.

Everyone feels the pull of these two forces, and in turbulent times what looms larger is anxiety. But human beings are designed to embrace change because we are the only creatures who can choose a creative life. A cat is content to eat, sleep, and lie in the sun. It was designed to be a creature of instinct, and very little free choice is allowed.

We have been taught through societal conditioning to fear change, and the teaching has been very effective so far. Since all of us have learned to approach life in terms of risk, threat, distrust, and worst-case scenarios, there is a lot to unlearn if we want a life filled with creative choices. The alternative is to become robotic, giving in to routine, habit, and conformity. They might feel like protection, but, in fact, their chief purpose is to hold off fear.

The way to unlearn fear of change begins by knowing a few things about how the whole process of change works. Here are some basic principles.

1. Change is embraced when it feels good to change.

This principle is important because it allows you to see change in terms of fulfillment. When you convince yourself that you hate change, what you are really affirming is that you've learned to connect change with a general sense of unease about a possible threat. The model for all imagined threats lies in the past with a bad experience you didn't overcome. At the emotional level, the prospect of change is already loaded with apprehension.

Without the shadow of the past, there is no reason to have any expectations. Few people would fear winning a million dollars or being offered a dream job. If you try to motivate yourself to change because it is good for you but not what you really desire, or for some other flawed reason—because other people don't like the way you are or you hate something about yourself (the most common things are your weight, your physical appearance, and your lack of success at work)—you will quickly discover that negative motivation doesn't work or, if it does, it leads to temporary results only.

2. Pain is a poor motivator for change.

Almost any other creature can be trained by offering a combination of reward and punishment, but this works very poorly with human beings. People stubbornly endure pain of every kind. Early on, punishing children creates a backlash, increasing their desire to disobey. It might seem mystical when the Buddha declares that pleasure is inevitably connected to pain, but that's a basic tenet of human nature. Swinging between the cycles of painful and pleasurable experiences, we are habituated to both and therefore are not motivated to change.

3. People behave in order to fit in.

Conformity is a powerful motivator for not changing. It works against creativity because to conform means never thinking for yourself. When you conform, you adapt your mind to secondhand values and received opinions. Most behaviors that make change frightening stem from the fear of not belonging, not fitting in.

Why is this idea of not fitting in so powerful? It acts like a survival mechanism. From prehistoric times onward, being part of the collective meant safety, while being separated from the tribe put your life in danger. What began as a physical need over time turned into a psychological habit. This doesn't make conformity any less powerful, but, with self-awareness, you realize that conformity doesn't really bring protection. It serves instead to block your individuality; therefore, it is adharmic, because the life you were meant to live is nothing if not individual.

4. Everyone takes the path of least resistance (or is tempted to).

We are all lured by inertia, the psychological force that keeps things the same, for no better reason than following the path of least resistance. Inertia can wear a positive face ("If it isn't broken, don't fix it"), but the sameness that inertia creates doesn't really satisfy. The main reason we follow the course of least resistance is that we've conditioned ourselves to go along in order to feel that we are accepted and belong. Feeling that you belong is a valid need, but once it has been achieved, there's no reason to make it the dominant feature of your existence.

5. Denial and resistance are the default responses in a difficult situation.

When most people run into trouble, face a looming crisis, exhibit possible disease symptoms, or feel overwhelmed, the same default response kicks in. They do nothing. They freeze. Denying to themselves that anything is wrong and resisting the best advice to get up and do something constitute automatic behavior. We all know this and have experienced what it is like to be stuck or paralyzed. Change can't occur under such circumstances, and therefore a new default response has to be learned.

6. All the behavior you think belongs to you was learned from someone else.

This is the insight that allows you to become who you really are. You are a creature of choice, change, and creative possibilities. The behavior and attitudes that stifle your true nature were learned somewhere along the line in your past. To unravel what the past has done to you is a long, exhausting, and ultimately futile task. The productive route is to embrace your true nature, which by itself will cause secondhand behavior to loosen its grip.

7. Your true nature lies in the present.

Everyone's life is overshadowed by the past. Whether you see this as a good thing or a bad thing doesn't obscure the fact that following the lessons of the past makes it impossible to live in the present. Only by living in the present can you find your true self. If you look even deeper, you realize that the present is just another name for the here and now. All the creative possibilities that seem unavailable have no secret hiding place. Existence is an infinite resource of possibilities, which is why human beings were designed to be creative—we have an enduring connection with the flow of creative intelligence.

These seven principles outline a path from fear of change to embracing life's creative possibilities. With a path to follow and a vision to inspire you, the way ahead is open. In practical terms, we are talking about finding wholeness. Change and non-change aren't opposites. They are the fundamentals of existence.

The trick is to adapt change and non-change to your inner life. What people lack isn't change, which is inescapable. They lack the foundation of non-change, which comes from self-awareness. Distracted by the constant activity of the mind, we overlook that consciousness itself doesn't change, not in its essence. The human brain offers a physical parallel to this. The functions performed by neurons are fixed by the laws of physics and chemistry. Yet within this rigid framework, the brain's activity embodies the mind's infinite creativity. By itself, a brain cell doesn't understand how this seeming incompatibility works. A cell understands nothing at all; it operates according to its genetic program.

The coordination of change and non-change happens in consciousness, which is the only place where understanding is possible. Consciousness remains unaffected by external flux. A lovely analogy for this comes from the Vedic tradition, which holds that the active mind is like the broken reflection of the moon on a choppy lake. No matter how turbulent the image, the moon is whole and intact.

This insight makes it much easier to embrace change and escape the fear of change. Your path unfolds, and over time a state of calm, restful, simple awareness becomes your default state. Secure in your awareness, you have the foundation for experiencing any change in the knowledge that you are infinitely adaptable. There is no higher or more satisfying way to live.

On Your Path

One of the main themes in this chapter centers on your attitude toward uncertainty. Now I want to bring that topic closer to home, offering practical steps you can take to evolve personally.

For anyone, the spiritual path is not entirely certain or entirely uncertain—it is a marriage of both. This only makes sense, because at this moment you are living with the life you know, while the evolutionary impulse of Dharma is urging you to step outside that life. What is popularly called an "aha" moment happens when something suddenly clicks, and the inner urge to evolve reveals something you didn't know before. This is a moment of self-awareness that brings inner growth and expansion. AI makes the same point in more detail.

What are the biggest "aha" experiences that transform people's lives?

ChatGPT

There's a myriad of "aha" moments that can transform someone's life. Here are a few:

Self-awareness: Realizing one's strengths, weaknesses, desires, and fears can be a profound turning point. It often leads to personal growth and a clearer path forward.

Empathy: Understanding and truly feeling another person's perspective can shift attitudes, relationships, and the way we interact with the world.

Failure: Embracing failure as a stepping stone rather than a stumbling block can transform one's approach to challenges and foster resilience.

Mindfulness: Being present and aware in the moment can change how someone experiences life, reducing stress and enhancing appreciation for the little things.

Learning from mentors or role models: Having a mentor or encountering a role model can offer insights that reshape values, ambitions, and approaches to life.

Gratitude: Realizing and appreciating what one has, can bring about a profound shift in perspective, fostering contentment and a positive outlook.

Forgiveness: Letting go of grudges and learning to forgive can be freeing, releasing the emotional burden, and allowing personal growth.

If your immediate response is that you've never had such life-changing moments, resist that reaction. Instead, ask yourself how you can encourage such experiences. Life-changing moments are like seeds that fall on fertile ground. Water them, and they spring to life.

The life you are leading right now is fertile ground waiting to be watered. Nourishing yourself is something you were designed for. To understand what that means, take a moment to reflect on each of the experiences listed above. In your reflection, ask three key questions:

Does this apply to me, even occasionally?

Do I value such experiences?

Do I want more experiences like this?

Let's consider the example of learning from mentors or role models. *Does this apply to you, even occasionally?* Everyone will say yes if they think back to childhood, because young children take their parents as role models, and, if they are fortunate, teachers in school continue the process. Superheroes are adolescent role models, largely because an adolescent can vicariously escape the awkward, self-conscious state of everyday existence by imagining themselves as a superhero. Adult role models reach for a higher ideal that is grounded in a great achievement or gift—an Einstein, Mozart, Lincoln, or Harriet Tubman.

Now the next question: *Do you value this experience?* On reflection, the answer is almost certainly yes. The ideals we see in our role models are internalized as ideal images of ourselves.

Do you want more experiences like this? That's the only tricky question. Like a poster on the wall to be gazed at with wishful thinking, the image you hold of your role model is static unless you are inspired to act on it. Passive admiration keeps you stuck where you are.

Realizing this, people try to have a dynamic relationship with their ideal. "What would Jesus do?" is a popular way of activating the New Testament to bring Jesus's teaching into the present moment. In the West, people are much less familiar with the role of the guru in Indian spiritual culture. Set aside for the moment any negative connotations this word has for you. Guru isn't a role that equates with cult leader. As mentioned earlier, in Sanskrit the roots of the word *guru* mean "dispeller of darkness."

This implies the highest kind of mentoring. A guru is meant to inspire you to live in the light, which is the ideal life in the Yoga tradition. What is imparted isn't knowledge about how to live or a set of teachings about morality, good and evil, closeness to God, or the like. The only thing a guru imparts is the quality of being that they radiate from a higher consciousness. In that sense, any bringer of light also dispels the darkness. A loving mother performs the role of guru by radiating love. A coach can inspire resolve, determination, and courage. A mentor in business can impart a sense of integrity to go along with material success.

More Self-Reflection Questions

You can reflect upon any "aha" experience to bring it closer to your life. In this way, change starts to acquire the quality of inspiration rather than threat. When asking such questions, your aim is to expand your awareness, not merely to get more information and advice.

Inspiration comes first. Ask ChatGPT to supply you with inspiring quotes on any topic—it has a virtually endless supply. For example:

Give me an inspiring Buddhist quotation on gratitude.

ChatGPT

Certainly! One beautiful Buddhist quote on gratitude is "Let us rise up and be thankful, for if we didn't learn a lot today, at least we learned a little, and if we didn't learn a little, at least we didn't get sick, and if we got sick, at least we didn't die; so, let us all be thankful."—Buddha

There is also inspiration to be found in affirmations.

Give me an affirmation about being more mindful.

ChatGPT

Absolutely. "I am present in each moment, embracing life with awareness and gratitude. I anchor myself in the now, allowing peace and clarity to flow through me. My mind is calm, my heart is open, and I am mindful of the beauty in every breath."

Starting your day with an inspiring quote or affirmation sets you on the lookout for the quality you want to increase in your life. You can go further and ask for practical suggestions. Here's a request about learning from failure.

I had an outcome at work that made me feel like a failure. I wish I could do things over again, but I can't. I want to turn my sense of failure into something more positive, something I can learn from. Give me one practical step I can take.

ChatGPT

When faced with a setback at work, it can be challenging to shift from feelings of failure to a positive learning experience. One practical step you can take is to conduct a "postmortem" or a retrospective analysis of the situation. When you are in a calmer mood, take time to reflect on what happened. What were the circumstances leading to the outcome? What factors contributed to the result?

A wise friend or a therapist might offer the same counsel, but with AI you feel secure in having control over how you pose your questions, your willingness to cooperate with AI's suggestions, and all the feelings that surround a disappointing setback. In fact, ChatGPT goes on to amplify what you can do moving forward.

Identify lessons: Identify specific lessons or insights from this experience. What did you learn about yourself, your approach, or the situation? Were there any skills or knowledge gaps that became apparent?

Actionable takeaways: Determine actionable takeaways from this analysis. What steps can you take to improve or prevent a similar situation in the future? It could involve acquiring new skills, adjusting your approach, seeking advice, or enhancing communication.

Implement changes: Implement the changes or improvements you've identified. This might involve seeking additional training, adjusting your workflow, seeking mentorship, or changing how you approach similar tasks.

Now you have in hand the template that will apply to every strategy that follows:

Inspiration

Affirmation

Self-reflection

Taking action

There's a great deal to say—and to be inspired by—as our journey continues, but, in a sense, learning to embrace change is the golden key. *Evolution is change, and change means choices.* When reduced to its essence, everything on the path stems from that one insight.

GUIDED MEDITATION
AI as Guru

The most profound role AI can play is the role of guru. Even at this early stage of AI's development, chatbots have access to the entire Indian spiritual tradition. It would offer the most help, however, to ask AI to distill this vast knowledge into practical exercises that you can use on the path.

I asked ChatGPT to cover seven areas of spiritual interest with guided meditation inspired by the seven strategies. Here is the first application, with the hope that you will invent your own prompts—this is only a starting point.

You are an expert in Vedic spirituality. Please create a guided meditation on the following topic: Building Awareness.

ChatGPT

Building Awareness

- **Setting:** Find a quiet, comfortable place to sit or lie down, where you won't be disturbed.

- **Focus:** Begin by focusing on your breath, noticing the natural rhythm of your inhalations and exhalations.

- **Visualization:** Imagine a radiant, glowing light at the center of your being, representing your inner awareness. With each breath, this light becomes brighter, expanding to fill your entire body.

- **Mantra:** Silently repeat the mantra *So Hum* (which means "I am That") with each inhale and exhale, aligning yourself with the universe.

- **Reflection:** Observe any sensations, thoughts, or emotions that arise, acknowledging them without judgment and returning your focus to your breath and the light within.

Put Your Ego on Notice

The problem of the ego is peculiar because it is very near and very far away at the same time. "I, me, and mine" are always with us, but we fail to notice that they have an agenda. This agenda doesn't consult what your dharma is, or even take it into consideration. What "I" wants right this moment is announced loud and clear, either as a desire you want to fulfill or a source of pain you want to avoid. There has never been a time when your ego didn't want something, but you are fooled into thinking that you're the one who is in charge.

The force of Dharma runs much deeper, which is necessary, because on the surface, an unending stream of wants, needs, and demands is self-perpetuating. There's a lot to come to terms with here, which is why most people aren't able to find a way past the ego's constant message—"more for me." There is also confusion between the words "ego" and "egotism." Unless you act very full of yourself, no one is ever likely to turn to you and say, "You have an ego problem." Yet one of the biggest "aha" moments on the path comes when you recognize that your ego is a problem. In some ways, it is *the* problem, because your ego's agenda stands in the way of your evolution.

Some traditions, especially in Buddhism, are rich in how they describe the subtlety of the psychological games that the ego plays, but such a complex treatment isn't necessary. We've touched already on "I" and "It" being your constant companions on the path (see page 14). "I" stands at the center of the story you have built around yourself since early childhood. By contrast, "It" stands for the silent impulse of pure awareness, the pull of the Higher Self, and the urge to evolve. Both forces are acting upon you all the time, in ways that are fundamentally incompatible.

"I" wants you to look out for number one.

"It" shows you how to go beyond selfishness.

"I" warns you that the world is a dangerous place.

"It" sees you as a privileged child of creation.

"I" is obsessed with the desire to maximize pleasure and minimize pain.

"It" shows you the eternal bliss that hides behind the mask of pleasure and pain.

As far as your ego is concerned, you would upset the applecart if you started listening to "It." Once it dawns on you that your ego is a problem, you have to ask yourself what to do about it. Here is where the tricky part begins. You and your ego are fused so tightly that it has become indispensable. One of the toughest spiritual exercises is to go an entire day without saying the word "I." (Even if you managed it, everyone around you would think you were being very strange.) A simpler exercise is more fruitful. Watch yourself and notice how many times you use the following verbal formulas:

I want X. I don't want Y.

I like this. I don't like that.

I really care. I don't care.

I agree. I disagree.

I feel good. I feel bad.

It doesn't matter whether you use these exact words or not. What all these verbal formulas have in common is that they keep "I" exactly where it wants to be, at the center of your existence. The voice of the ego is constantly playing in your head, so it has become second nature. The tricky part is how to hear "It" without "I" filtering the message and twisting it around. The ego is like a perverse switchboard operator. Instead of putting a telephone call through, the operator starts telling you what you want to hear.

On the dharmic path, you need a strategy for putting your ego on hold so that the impulse of evolution registers with you. Otherwise, "I" is just putting on another disguise, the disguise of the spiritual seeker, the virtuous do-gooder, the selfless saint, or the wise knower who has figured out how life really works.

AI can give us a condensed summary of all this.

For spiritual seekers, the ego is seen as an obstacle. Tell me why in 50 to 75 words.

ChatGPT

In spiritual seeking, the ego is viewed as a barrier because it reinforces attachment to the self, fostering separation, desires, and fears. It obscures the true nature of existence, hindering the quest for deeper connections, empathy, and enlightenment. By transcending the ego's limitations, seekers aim to attain a more profound understanding of the self and interconnectedness with the universe.

Whether this answer sounds alien or familiar depends upon your background, particularly your religious background. The Judeo-Christian tradition doesn't explicitly use the word *ego,* shunting the issue into questions about sin and redemption. Instead of the ego being the source of temptation, the sinner is, going back to Adam and Eve eating the fruit of the tree of knowledge of good and evil. Instead of "transcending the demands of the ego," the Christian ideal is to be in the world but not of it, which says the same thing. The cherished goal of redemption is to leave the sinner behind so that the purified soul reaches Heaven.

By contrast, Eastern traditions are more direct, as a question put to AI reveals.

Briefly define the Buddhist concept of ego death.

ChatGPT

"Ego death" in Buddhism refers to the dissolution of the illusory sense of a permanent, independent self. It's the realization that the self is impermanent, constantly changing, and lacks inherent existence. Through practices like meditation, one aims to transcend the ego, recognizing its transient nature and achieving a state of liberation from attachment and suffering.

In the modern secular world, many people don't feel the influence, much less the pressure, of a religious background. Good and bad consequences flow from this, which one can argue over endlessly. But one psychological advantage is undeniable if you didn't inherit a strong religious tradition. Religions impose secondhand beliefs on the devout, and if this burden is lifted, the problem of the ego can be seen in a clearer light.

The Riddle of Desire

The reason that you want anything is because your ego wants it. A constant stream of desires, large and small, runs through the mind. Desires aren't merely distractions. They are the core of everyday life. The ego's refrain of "I want this" and "I don't want that" is inescapable. The spiritual implications are briefly summarized by AI.

In Eastern traditions, what is the connection between the ego and desire? Using simple language, tell me in 50 to 75 words.

ChatGPT

In Eastern traditions like Buddhism and Hinduism, the ego and desire are closely linked. The ego thrives on desires, making us feel separate and needy. Desires, whether for things or ideas, create a cycle of longing and dissatisfaction, causing suffering. By letting go of desires, these traditions teach that we weaken the ego's hold and find peace, aiming for liberation from the constant chase after fulfillment.

It is very clear, and has been for many centuries, that desire is considered a form of bondage in Eastern traditions. (The closest Christian equivalent is the seven deadly sins—greed, lust, and gluttony are connected with excessive appetite. More indirect are pride, sloth, envy, and wrath, but they can easily be translated into the desire to be better than others, the desire to be lazy, the envious desire to have what someone else has, and so on.)

Getting free from the clutches of desire reveals a riddle that seems impossible to solve. You can't actually live without desire, beginning with the basics of survival (the desire for food, home, and shelter from the cold) and extending to the desire for love, sexual gratification, affection, respect, and self-esteem. To the ego, fulfilling these desires makes life worth living. So the problem isn't that people greedily want to amass more money, possessions, status, and power. The problem is how to relate to desire in the first place.

If you are honest with yourself, the prospect of renouncing desire holds no appeal, even if you consider yourself deeply spiritual. You can seek a mountain retreat or forest ashram, but desire will follow you doggedly. Wanting to be close to God is a desire, too, and so is wanting enlightenment. The ego views everything as an either/or choice, including the either/or of being worldly or spiritual. You have automatically adopted this viewpoint through habit and indoctrination. What has been missing is a different perspective.

Adopting an evolutionary perspective is essential to progress beyond your existing patterns of behavior. We can specify what this new perspective should be like. It shouldn't ask you either to struggle against your desires or to give in to them. It should be satisfying to follow. It should give you a higher view of yourself.

Here are four principles that outline an evolutionary approach, based on being more aware of how desire can work for you on the path:

> Be aware of the cycle of desire.
>
> Examine your impulse control.
>
> Measure the fulfillment of desire by the joy it brings you.
>
> Give priority to your highest aspirations.

These four precepts apply to any desire you want to address, although it is only reasonable to respect the difference between craving the latest iPhone model and desperately wanting an opioid fix. The grip of addiction or a desire that has become self-defeating, perhaps even self-destructive, calls for professional intervention. But there is still a wide range of desires that can be addressed, and you need to if you want to listen to "It" rather than "I."

Be Aware of the Cycle of Desire

The lure of any desire begins with its immediacy. Desires want you to act now. Delay is frustrating and only serves to double the intensity of the desire. This is part of the ego's power and shouldn't be related to adolescent hormones in the case of sexual desire or primitive impulses like anger, envy, male dominance over other males, etc. The real issue is a failure to look beyond the moment. Once you do, you become aware that "I want this now" is just the first step in a predictable cycle.

First comes the immediate feeling of wanting something.

You act in pursuit of the thing you want.

You get it and are satisfied.

The satisfaction wanes.

Once this happens, room is made for the cycle to begin all over again. The cycle of desire is natural and operates every day in how we eat, for example. You feel hungry; you get something to eat; it feels satisfying; satiation gradually wears off; and you find that you are hungry again. Being aware of this basic rhythm is helpful in the psychological treatment of obesity, where the person who habitually overeats is asked to wait until he feels hungry. If the cycle is curtailed, eating becomes a habit divorced from its natural cycle (for instance, when somebody "eats their emotion" as a quick fix for feeling bad). Then no matter how much you eat, there is no satisfaction that has any lasting effect. That's when people say, "I can't believe I just ate that." Nothing good was attained.

Many desires are turned into similar problems as overeating, including sexual desire, the urge to win, wanting more money, and seeking control. Hypercompetitive people aren't satisfied by winning. It has become a fix that is practically meaningless but is fueled by a compulsive drive. (Next time you watch a competitive sport, whether it is football, soccer, baseball, or tennis, notice the expression on the face of the winners. It is more often angry than happy, reflecting a warrior mentality that is motivated by crushing an enemy rather than rejoicing in the victory.)

With this in mind, begin to anticipate where the cycle of desire is taking you. You are replacing a reflex—responding to the immediate urgency of a desire—with self-awareness.

Examine Your Impulse Control

Most eruptions of violence, whether it concerns a fight on the playground, domestic abuse, or an incident of road rage, involve the loss of control. Being a mature adult means having a measure of control over your emotions, but the issue goes deeper than that. Desires arise as impulses, which in young children hold sway until they learn through their parents that there is a second stage, where the impulse is examined. The mind goes through reasons for accepting or rejecting the impulse.

Some reasons are primitive. *I might get caught. I might be punished. Mommy and Daddy will get mad at me.* These are familiar thoughts and patterns from early childhood.

At some point, higher considerations enter into the picture as we grow older. *I might feel guilty. This didn't work out so well the last time. I don't want to look bad.*

When we are mature adults, moral reasoning and a personal code govern our actions. *This is wrong. I will regret it tomorrow if I give in. I couldn't live with the guilt. I have a responsibility to others for what I do.*

However, such neat categories don't apply in real life. As adults, we hear childish, immature voices urging us to go ahead despite the warning mature voice that knows how badly things might turn out. A stronger sense of impulse control would save people from rashly saying things in a relationship that can never be taken back. In turn, fewer of us might not cheat in business, lie on tax returns, or cheat on a partner or spouse.

Unfortunately, maturity isn't the same as self-awareness. Mature people act from the level of ego all the time. Fighting against your desires and trying to rein in your impulses aren't good long-term tactics. Desire only grows stronger when it is suppressed. The answer is to be aware that desires are ultimately about feeling fulfilled, which can only become lasting through closer contact with the source of all fulfillment in pure awareness.

Measure the Fulfillment of Desire by the Joy It Brings You

Desire brings happiness in one of two ways—either at the start or at the end. If you feel joy by having a desire that enhances your life, that is more secure than waiting for the eventual outcome. You are motivated from the start. Waiting for a result is a fickle proposition by comparison. Here's a short list of blissful desires that will enhance your life:

Wanting the best for someone else

Desiring peace and an end to violence

Helping to heal a fractured friendship

Wanting your children to be happy

Seeing the best in others

Lowering the level of stress around you

Wanting to be of service

It is entirely possible to have any of these desires without feeling joy. You can want peace because the prospect of war and violence makes you feel despair. You can decide to reconcile with a friend who is mad at you while still believing that you are right and the friend is wrong. Spiritually, the trajectory of any desire is directed by the motivation that goes into it at the start. There's all the difference in the world between feeling overjoyed at the prospect of having a baby and wanting to have a baby because your marriage is on the rocks. Using pregnancy as a fix for relationship problems never works. The strain only becomes worse once the baby arrives and caring for a newborn brings up new stresses.

One of the tragic aspects of addiction is that the drug of choice, which began by bringing a jolt of pleasure, eventually stops delivering that effect. The craving for the drug continues (we can amplify the meaning of "drug" to include sex, food, or the overwhelming need to win), but any so-called benefit has been extinguished. What is left is pure compulsion. The drive for the next fix is everything.

As dire as addiction is, normal desires ultimately pose the same drawback of diminishing returns. The goals that society approves of—acquiring money, success, status, power, and possessions—become an end in themselves. Stories filter through the media of grotesque extreme behavior. For example, after Ferdinand Marcos was deposed as dictator of the Philippines in 1986, press reports took note of Imelda Marcos's lavish wardrobe, said to include 15 mink coats, 508 gowns, 888 handbags, and 3,000 pairs of shoes.

What this underscores is the disconnect between wanting something and getting it. In between, there is a lack of fulfillment. You can repeat your behavior over and over, wanting more and getting more, but the missing element only becomes more glaring. Whatever power Imelda Marcos accrued through unlimited acquisition, she was a puppet of desire.

On the path, you measure fulfillment by the joy your desire brings you. A joyful desire is well on its way to a joyful conclusion. Being aware of this can make a great difference.

Give Priority to Your Highest Aspirations

By now, this idea will sound familiar to you. The worth of being on the path is to make room for love, compassion, empathy, creativity, and the other highest impulses that arise from pure awareness. But it is easy to underestimate how mechanically we repeat the same pattern of desire without sorting those desires into the best priorities. One aspect of deeper wisdom is that intention has hidden power. Consciousness is an infinite field that everyone shares. If the majority of people inject prejudice, ill will, and hatred into the field, those impulses register, and what bounces back is a reflection of what went in. The adage about writing computer software, "Garbage in, garbage out," is entirely applicable to your intentions.

In situations that became historical tragedies—antisemitism in Germany, apartheid in South Africa, racism in the American South— terrible actions were taken by a minority of the population. The "good people" passively went along or turned their backs. They attempted to keep up the appearance of normal life by denying the rot at the core of their society. Passivity and denial feel like safe havens. But Ralph Waldo Emerson's famous adage exposes the truth: "The only thing necessary for the triumph of evil is for good men to do nothing."

On the path, you are aware of your intentions as if they were actions—thinking, after all, is a mental action. Through awareness you create an opening for your highest aspirations in a simple way: You pause when you detect a damaging intention and stop favoring it. Here are the kinds of damaging intention all of us might easily overlook or even promote:

Wanting an enemy punished

Considering violence to be necessary

Spreading juicy or malicious gossip

Wanting to reveal a secret told in confidence

Indulging in revenge fantasies

Clinging to bad memories

Reliving old traumas

Harboring resentment in your relationship

Being tempted to do what you know to be wrong

Excusing yourself from doing what you know to be right

Letting guilt and shame fester

I don't mean for this list to be daunting. It serves as a reminder that the active mind is haphazard, restless, and unpredictable. It has no way of governing itself, and if you push to impose control, the forbidden thoughts seek cover for a while, waiting for the next opportunity to rise again.

What you pay attention to grows. By pausing as soon as you notice a damaging thought, you stop giving it more fuel. If you have a moment, you might look at the thought and reflect on it. Why do you have a revenge fantasy? What good is it really doing? How does it make you feel? The more you reflect, the more you drain away the power of a damaging intention. As a secondary benefit, you feel good about quashing bad intentions. Such interventions might seem small at first, but over time you will find that you trust your desires to be life-supporting. That's a major advance on the evolutionary path.

On Your Path

To put it simply, letting your ego be the boss keeps you enthralled in the cycle of desire. But there is a way out. Taking power away from your ego breaks the bonds and allows desire, which is a natural part of life, to return to its role as your ally on the path. AI gives very clear guidance here.

If I am on the spiritual path, how can desire be my ally instead of an obstacle?

ChatGPT

Desire can indeed be a powerful force on the spiritual path, and whether it is an ally or an obstacle depends on how you approach and manage it. Here are some ways to make desire your ally on your spiritual journey:

Practice self-awareness: Start by understanding your desires. What are you truly longing for? Is it material wealth, emotional fulfillment, or spiritual growth? Self-awareness can help you differentiate between fleeting, ego-driven desires and deeper, soulful yearnings.

Align with higher values: Channel your desires toward spiritual growth and self-improvement. Align your desires with values such as love, compassion, wisdom, and inner peace. When your desires are in harmony with these values, they can become powerful tools for personal transformation.

Set intentions: Instead of chasing after desires aimlessly, set clear intentions. Define what you want to achieve on your spiritual path and why it matters to you. Intentions can help you stay focused and make better choices.

Detach from outcomes: One of the main obstacles desire presents on the spiritual path is attachment to outcomes. Practice detachment by accepting that the results may not always match your desires. Trust in the journey and the lessons it brings.

Practice mindfulness: Cultivate mindfulness to observe your desires without judgment. Mindfulness helps you become aware of your desires as they arise and allows you to respond consciously rather than react impulsively.

Practice gratitude: Gratitude can help shift your focus from what you lack to what you have. Regularly express gratitude for the blessings in your life, which can reduce the intensity of your desire for more.

Surrender and trust: Trust the universe or your spiritual beliefs to provide what is truly necessary for your growth. Surrendering to the flow of life and trusting in divine wisdom can ease the grip of desire on your consciousness.

Use desire as a motivator: Instead of letting desire hold you back, use it as a motivator for personal growth. Let your desires inspire you to work on yourself, develop new skills, and make positive changes in your life.

Seek guidance and community: Connect with like-minded individuals and seek guidance from spiritual teachers or mentors who have walked the path before you. Their wisdom can help you navigate the challenges of desire on your journey.

Practice self-discipline: While desire can be a powerful ally, it can also lead to distraction and indulgence. Develop self-discipline to manage your desires and stay on track with your spiritual goals.

Remember that desire itself is not inherently good or bad; it's how you relate to and manage your desires that determines whether they become allies or obstacles on your spiritual path. By aligning your desires with your higher purpose and practicing mindfulness and self-awareness, you can harness their energy to propel you forward on your spiritual journey.

These points are consistent with the vision being unfolded in this book, and it is reassuring to see it being affirmed. You can personalize each point by asking AI to give you an update every day, choosing the point you'd like to focus on. Here's an example of setting up a dialogue rather than settling for a blanket answer.

Let's take one point you made—using desire as a motivator. In simple language, give me one quote that will inspire me today.

ChatGPT

"Let your desires fuel your determination, and watch your dreams take flight."

On the same theme, give me a single affirmation.

ChatGPT

"I harness the power of my desires to fuel my passion and drive towards my goals."

On the same theme, give me a question to reflect upon.

ChatGPT

"What is the burning desire within me that, if pursued with unwavering determination, could lead me to my most meaningful goals and dreams?"

Engaging in a conversation that continues for as long as you like is always an asset with AI, but your aim should be to find a response that inspires you or gives you a new insight. Here's an instance of what I mean.

Give me some wise words about desire. Cite a specific spiritual teacher, guide, or authority.

ChatGPT

Here are some wise words about desire from the renowned spiritual teacher and philosopher Jiddu Krishnamurti:
   "Desire is the starting point of all achievement, not a hope, not a wish, but a keen pulsating desire which transcends everything."

Krishnamurti emphasizes the importance of a strong and focused desire as a driving force for achieving one's goals and aspirations.

That's beautiful. Can you give me wise words from another spiritual guide or teacher?

ChatGPT

Certainly! Here are wise words about desire from the Indian spiritual teacher and leader Mahatma Gandhi:

"Satisfaction lies in the effort, not in the attainment. Full effort is full victory."

Gandhi's words remind us that the process of pursuing our desires with sincere effort and dedication is where true satisfaction and victory are found, rather than solely in the attainment of our goals.

As you can see, going beyond useful information to a level of personal inspiration is just as convenient. You are taking time in a meaningful way to put your ego on hold, making room for an evolutionary message to come through.

GUIDED MEDITATION
AI as Guru

Following the pattern of asking AI to create original guided meditations in the role of guru (see page 84), here is what ChatGPT devised for Loving-Kindness. Feel free to invent your own prompts once you use this one.

You are an expert in Vedic spirituality. Please create a guided meditation on the following topic: Loving-Kindness.

ChatGPT

Loving-Kindness (Metta)

- **Setting:** Sit comfortably with your eyes closed, grounding yourself in the present moment.

- **Heart space:** Bring your attention to your heart, imagining it filled with a warm, gentle light.

- **Mantra:** Silently repeat *Om Sarva Mangalam* or it might be easier to repeat phrases of loving-kindness toward yourself, such as "May I be happy, may I be healthy, may I be at peace."

- **Expansion:** Gradually extend these wishes outward, first to loved ones, then to acquaintances, and finally to all beings everywhere, envisioning waves of loving energy radiating from your heart.

Bend and Bounce Back

The reason that we put up with stress every day is that we want to put up with it. At first blush, you might not agree with that statement. None of us wants anxiety, worry, or pressure in our lives. However, modern society has struck a deal where stress is the price we pay for certain rewards: the salary a well-paying job pays, the security of having a family even though the family might be severely dysfunctional, the comfort of a lasting relationship even though nothing is more difficult than living with another person. All these benefits of life come with a certain number of stressors. If you aren't aware of those stressors, the price paid rises higher over time, and the rewards don't balance it out. Is it worth making a lot of money if the stress involved wrecks your health and family?

The typical advice is to learn better coping skills, and it is good advice. Society, in general, doesn't teach us how to cope; it is something we must teach ourselves. On the path, however, we can develop a strategy that goes much further than just coping. It begins with a saying that might or might not come from Confucius but in any case is very old: "The green reed that bends in the wind is stronger than the mighty oak that breaks in a storm."

Stripped of its poetic veneer, this saying speaks to resilience—the capacity to adapt and recover in the face of life's challenges. We were designed to be resilient; rigidity is a learned defense. Rigid people are pretending to be strong to hide from how vulnerable and threatened they feel inside. If you are afraid enough, you constrict your awareness like a snail pulling into its shell. Fear is the enemy of evolution, which requires expanded awareness.

AI can help to bring this notion closer to home.

People speak about expanded awareness. To make that real, I'd need to know what it feels like. Tell me briefly in simple language.

Gemini

Instead of a narrow spotlight, your awareness becomes a wide, gentle floodlight. You're not just focused on one thing at a time, but holding the whole scene in your mind, noticing everything that unfolds. Here's what that might feel like:

- **Sensory richness:** You're more aware of subtle details—the warmth of the sun, the scent of coffee, the faint hum of a bee.

- **Inner calm:** You're not constantly pulled by thoughts or emotions. There's a quiet space within, even amidst activity.

- **Presence in the moment:** You're not lost in the past or future, but fully engaged in the present, savoring each breath.

- **Non-judgmental observation:** You see things clearly, without immediately labeling them as good or bad. You simply observe.

- **Openness and curiosity:** You feel a gentle interest in everything around you, a desire to learn and understand.

It's like you've stepped back and gained a wider perspective on life, appreciating the interconnectedness of everything.

This is just a glimpse. Expanded awareness is a personal journey, and its specific sensations may vary. But hopefully, this gives you a basic idea of what it might feel like to "floodlight" your awareness and experience the world in a richer, more present way.

That's a beautiful description, but what brings it within reach, practically speaking, is learning to be more resilient. You have to test your boundaries to make them expand. The experience isn't like breaking down a wall but leaning against it until it moves. Plaster and stone walls won't give just by leaning against them; however, mind–made walls do. I once heard an optimistic saying that makes a good guiding principle: "Souls don't break, they bounce."

When I was researching how people age for a book I was writing, I discovered that aging is a mysterious process. No two people age alike, even identical twins. There is no disease known as aging, and aging isn't fatal. We die because a major system or organ fails. At the moment that happens, the vast majority of the body's cells are working normally, perhaps as much as 99 percent if there is a fatal heart attack or stroke in an otherwise healthy body. But one confronts a unique mystery when it comes to people who live to be a hundred.

Centenarians are rare—in the United States your chances of living to be a hundred are around 1 in 5,000, and 85 percent of centenarians are women. For everyone, longevity depends on a host of factors. Having good genes is a start but not definitive. Other influences play a major role. In society today, people smoke cigarettes less than older genera- tions did, although alcohol consumption per person is about the same as it was in 1990. Modern medicine saves many more stroke victims who would have died in the past. The list of external factors goes on, but what about on the inside?

I discovered that people who live to be a hundred are in many ways models of resilience, which gets expressed in positive personality traits. Here is an AI description of their general outlook on life.

Gemini

- **Adaptability and flexibility:** Centenarians have witnessed significant historical and social changes. Their ability to adapt and learn new things might contribute to resilience and well-being.

- **Positive outlook and optimism:** Many studies show a tendency toward positive emotions, gratitude, and a sense of purpose, which might contribute to mental well-being and longevity.

- **Emotional regulation skills:** Centenarians might have developed effective strategies for managing stress and negative emotions, contributing to emotional stability and resilience.

A person can be lucky in the personality they were born with as well as their genes, but something deeper is going on here. As a very select group of society, centenarians seem more evolved. Their resilience, along with their optimism, is a product of expanded awareness, not simply a series of random factors. Life didn't teach them—many old people become resigned, bitter, or passive the longer they live. Other things happened "in here" among centenarians. Typically they grow more conscientious than other people; they place more value on meaningful social connections; they are more in control of their emotions; and many live with a deep sense of purpose.

It is unreasonable to declare that anyone who reaches a hundred was necessarily on the path. The concept of a path didn't exist in the West a century ago, except in religious terms. What happened instead is the natural process of awareness expanding when someone is attuned. To stay attuned "in here" decade after decade only happens if you are resilient.

Learning As You Go

Instead of leaving the process to chance, you can consciously work on becoming more resilient. To begin with, you must have a grasp on what it feels like to be rigid instead. This puts you in touch with what needs to become more flexible. When I asked Gemini for a simple index of personal traits that separate rigid and flexible personalities, it came up with the following:

Rigidity

I prefer well-defined routines and schedules.

I dislike changes in plans or unexpected situations.

I find it difficult to adapt to new information or perspectives.

I tend to stick to my established opinions and beliefs even when presented with contrary evidence.

I struggle to cope with ambiguity or uncertainty.

Flexibility

I am open to trying new things and exploring different possibilities.

I can easily adjust my plans and routines when necessary.

I am comfortable changing my mind based on new information or understanding.

I embrace diverse perspectives and can adapt my communication style to different audiences.

I find uncertainty exciting and see it as an opportunity for growth.

These general descriptions show you the direction you should be heading in, toward more flexibility. Self-awareness begins with noticing. Whenever you notice yourself being inflexible, pause and reconsider your response. Inflexibility is associated with feeling tight and tense. You are likely to be resisting someone or something. You hear yourself saying words you've repeated many times before. You feel that you are right and someone else is wrong.

Those signs are good indicators that you need to be more flexible. When you pause to reconsider what you are doing or saying, take a deep breath and let a new impulse come to your notice. If you are feeling tight and tense, nothing might appear. You can still back off, retract your resistance, and wait for a better frame of mind. Resilience is learned. Each time you notice your inflexibility and back away, you are softening the mind-made boundaries that constrict awareness. The evolutionary process is working.

On Your Path

With AI there is always the option of opening a conversation with a chatbot about a personal issue, as we've done several times already. Now I'd like to introduce an alternative, using a hypothetical friend who stands in for you.

This is a familiar tactic in real-life conversations: "Let's say I have a hypothetical friend who has a crush on someone." We resort to this tactic to avoid embarrassment. But you can also get insight about yourself this way. This is applicable to rigidity because "my way" doesn't like to be challenged and quickly grows defensive.

We will put the burden on someone else who is rigid so that you aren't the target. ChatGPT has a lot to say here and to keep its response from feeling like an impersonal laundry list, I'll comment on each point, turning the focus on you, the questioner, instead of your hypothetical friend. Let's jump right in.

I'd like to talk with you about my friend Bea. She is 62 and very bright, having worked for years as a paralegal. Our friendship is under strain because she is getting more and more opinionated, stubborn, and set in her ways.

ChatGPT

It's essential to approach the situation with understanding and empathy. Here are some steps you can consider taking to address the strain on your friendship with Bea:

1. **Reflect on your own feelings:** Take some time to think about why Bea's behavior is bothering you. Are her opinions and stubbornness the primary issue, or are there underlying factors that are contributing to the strain in your friendship?

We don't tend to reflect but instead to react. A stubborn friend is irritating, and if things get worse, it's hard not to react with anger and impatience. But turn the light on yourself. When you are in a stubborn mood, don't you feel justified? The voice that says, "I'm not giving in," is automatic. But you can turn inward and ask why. There is always an opening for self-reflection.

Why do I feel like I can't give in?

What makes me feel that I have to be right?

Is being rigid worth the price of alienating others?

Isn't there a grain of fear behind my stubbornness?

2. **Open communication:** Honest and open communication is key to resolving any issues in a friendship. Express your concerns to Bea in a non-confrontational way. Use "I" statements to describe your feelings and avoid making accusations. For example, you could say, "I've noticed that our conversations have become more challenging for me recently, and I feel like I can't express my thoughts and opinions freely. I value our friendship and want to understand how we can improve our communication."

If this advice seems reasonable, take the next step and ask how well you are communicating with yourself. Every voice in your head is talking to you from a fragment or microchip of your awareness. Do you talk back? When fear, resentment, envy, hostility, and insecurity speak, do you simply listen? Most people do just that. They might struggle with a negative reaction, stuff the voice down out of sight, turn their back, or go into denial. These reactions aren't the same as communicating.

When you hear a voice inside yourself, which might be like words in your head or just a familiar impulse, you have the option of having your say. These voices come from the divided self. They represent old, conditioned reactions; holdovers from childhood; and emotions we have been repressing. It takes awareness to quell them and find a better place emotionally.

Here are replies you can pose if you really want to communicate with these fragmentary parts of yourself:

Thanks for your input, but I don't need you right now.

I've heard this before.

Why are you talking to me now?

Are you the me of today or a remnant of the past?

What do you want?

Tell me how I can make peace with you.

3. **Listen actively:** When you talk to Bea, make sure to listen actively to her perspective as well. Try to understand why she may have become more opinionated and set in her ways. Sometimes, age-related changes or personal experiences can influence a person's behavior and beliefs.

The issue here is empathy. When someone close to you acts out in an overt way, all too often you find your defenses going up. You don't want to hear it. You are fed up. It's the same old thing over and over. Tuning someone out becomes more automatic the less you agree with them. Tuning in, on the other hand, requires empathy.

Like most complex human qualities, empathy has more than one aspect. It requires you to be open, non-judgmental, accepting to at least a degree, and emotionally sympathetic. Turning inward, how many aspects of empathy do you show to yourself? Whenever you are self-critical, you are accepting the division inside yourself that fosters blame, regret, self-doubt, and lack of acceptance.

Logically, who is criticizing whom? There's only you in there. The judgmental part, the "me" that is hard on you, is a fiction, a kind of Mr. Hyde to Dr. Jekyll. Those two personas existed in the same person, and they thrived in separation because neither had anything to do with the other.

Being resilient means being easy on yourself the way you'd be easy on someone you love. Getting there is a process, like everything else. But your aim is clear from the start—bring yourself into the state of wholeness that doesn't buy into the divided self. For example, when you hear yourself thinking *I hate myself,* even casually, pause and respond with *No, I don't. This is just an attitude I'm experiencing in the moment. The feeling will pass.*

Countering the impulse to be hard on yourself will feel unusual at first, but it falls under the category of communicating with yourself. The internal dialogue running through your head is old business trying to interfere with new business. It's the past trying to leach energy from the present. Ideally, there is no internal dialogue, because if you are in the present, you respond here and now, not with leftovers from the past.

Here's a sample of the kinds of thoughts you can firmly reject out of hand:

I'm not good enough (smart enough, pretty enough, thin enough, etc.).

I wish I were someone else.

I never learn.

Nothing ever goes my way.

If I didn't have bad luck, I'd have no luck at all.

If I had more money, everything would get better.

No one's on my side.

I don't really like myself.

Each of these self-defeating statements can be replaced by a thought that shows empathy for yourself and also contains more truth:

I'm good enough. Who says I have to be perfect?

Things do go my way sometimes.

People care for me when I take the time and effort to relate to them.

I might not like myself at this moment, but it will pass.

Who appointed me as judge and jury against myself?

How do I make things any better by being so hard on myself?

It is a very good practice, whenever you hear a voice deprecating or criticizing you in your head, to pause and replace the criticism with a positive statement. Your inner critic is repeating the past; your response is new and immediate, rejecting statements that are rigid holdovers from a time that no longer exists.

4. **Find common ground:** Emphasize the aspects of your friendship that you both cherish and have in common. Focusing on shared interests or memories can help reinforce your connection and remind both of you why your friendship is valuable.

This point is all about reconnecting. Reconnecting with a friend by sharing common interests and memories isn't the same as reconnecting with yourself. Society offers no good models for that. The reason that there is a disconnect "in here" goes back to the divided self, which creates an internal dialogue where none needs to exist.

Your mind is unified at the source. A thought or impulse rises, expecting that you will notice it without distortion. But as it moves toward your attention, there are filters that intervene or distort the message. These filters are a haze of memories, habits, beliefs, prejudices, secondhand opinions, and social conditioning. Some are safeguards, which we've discussed already as the ego's attempt to make life predictable and seemingly safe. Other elements are a hovering cloud of assumptions you've picked up randomly over the years.

Like haze on a wintry road that blocks your view, mental haze is confusing. Messages from your deeper self might not come through at all, but, generally, you feel undecided. Pure love, compassion, altruism, generosity, and empathy are hard to connect with, much less act upon. We hesitate, pull back, and have second thoughts.

Part of resilience is clarity, and to create clarity, you can't spend your time trying to sort out mental haze. The tactic that works is to always act on your highest impulse. Then it doesn't matter that you hear conflicting or confused messages inside. Do the loving, generous, compassionate, empathic, altruistic thing, even though the impulse isn't pure. It will be pure enough.

5. **Respect differences:** It's natural for friends to have differing opinions and beliefs. Remember that diversity

> in thought can be enriching and that you don't have to agree on everything. Try to respect Bea's viewpoints, even if you don't share them.

The gist here echoes Ralph Waldo Emerson's famous quote "A foolish consistency is the hobgoblin of little minds." Consistency isn't a positive trait in and of itself. If you insist on repeating the same reactions, the same opinions, and even the same words, by definition you are being rigid. But look at how often we keep on repeating behavior that got us nowhere in the first place.

That is what happens in relationships that circle back to the same arguments time and again, with neither partner finding a way to break the cycle. The same happens "in here." Take a problem like weight loss, which millions of people struggle with. There is a war going on inside between the impulse to overeat and the desire to "be good." If this war could be settled, it would have been long ago. But the clash of differences gets repeated incessantly.

Abandon repetition when it doesn't work. Give up consistency for the sake of consistency. When you feel inner conflict, the solution lies at a deeper level than the problem, and the point of personal evolution is to find this deeper level so that it can work for you.

> 6. **Set boundaries:** If there are specific topics or behaviors that are causing friction, consider setting boundaries in your conversations. Let Bea know which subjects are off-limits or how you'd like to approach contentious topics in a more constructive manner.

On the path the issue of boundaries is tricky. The ego imposes mind-made boundaries born of fear and insecurity. Your source in pure awareness is unbounded. As you evolve, you don't attack the boundaries that keep your mind closed and constricted. Attack only makes their resistance stronger, and you are faced with pushback.

What works is to allow awareness to soften and melt away mind-made boundaries. Yet during this process, you will still be living with all kinds of boundaries, and some are beneficial. Among these are the boundaries of

Respecting other people's space and having them respect yours

Not intruding with advice when it isn't welcome

Rejecting offensive speech, especially when it carries some kind of prejudice

Refusing to engage in wrongdoing

Resisting the temptation to lie or cheat

Maintaining your self-esteem

Speaking your truth

These boundaries are not meant to be rigid. You can be tolerant and flexible while still holding your ground.

---

7. **Seek compromise:** If disagreements arise, look for compromises or ways to agree to disagree without damaging your friendship. Finding middle ground or acknowledging that you both have valid perspectives can help ease tension.

When and how to compromise when it comes to relationships poses a daily challenge. Sometimes there's no common ground between "my way" and "your way," and the more rigidly you stick to your way largely determines how well you relate in the social world as well as at home behind closed doors.

Turning the same challenge inward takes most people into uncharted waters. To compromise with yourself sounds peculiar, although we do it all the time. In a restaurant, there is the compromise between no dessert and the most fattening chocolate temptation on the menu. Choosing a movie everyone will like, leaving early enough to be on time, packing enough for the beach but not too much—these situations crop up continually.

Other compromises call for negotiation, which is the point here. Being resilient doesn't mean you sacrifice your principles or go back on promises you made to yourself. It means negotiating between two impulses that are both valuable and speak for themselves. Conflict resolution is the only way to avoid the outbreak of war and also the only way to get out of a war without both sides annihilating each other (painful examples face the world as I am writing these words).

Every external conflict reflects an internal one. Take any hotbutton issue like gender equality, racial justice, abortion, or immigration. If you take one side of the issue and see no room for compromise, you are being rigid. If you can see both sides and weigh their pros and cons, you are being resilient. Only after this process of weighing and balancing takes place are you in a position to act. But personal evolution goes further than even that. There is a level of awareness where the solution to any problem emerges clearly and it isn't arrived at by debate, weighing pros and cons, or negotiations. On the path, you move in the direction of connecting to the level of the solution, which begins by not being immersed in the level of the problem.

---

8. **Spend quality time together:** Make an effort to spend time with Bea in non-contentious situations. Engaging in enjoyable activities or simply having lighthearted conversations can help strengthen your bond.

Many people find it easier to spend quality time with someone else than with themselves. Much of this can be traced to stress and the epidemic of central nervous system overload (see page 56). It is easier to rein in a slow pony than a racehorse. The same applies to the mind. The more active, stressed, and overloaded your mind is, the less space there is for quality time with yourself.

Here, quality time begins with downtime. You need to create a few moments several times a day when you do nothing but sit quietly. If you can, merge downtime with meditation taken in short bursts throughout the day. Then there is creative time, devoted to a hobby or an art that brings you unqualified pleasure. The more you consider it, quality time implies intimacy, communion, peace and quiet, and private communication. You'd want all of these things with someone you love. They should be extended to yourself as well.

---

9. **Give it time:** Changing long-standing behaviors and attitudes can be challenging. Be patient and give Bea some time to adjust to the idea of compromising or being more open-minded.

Tension is an unnatural state for the body, and given the first op-portunity, your muscles will relax and tension will be released through sleep. Time isn't doing this. The body's natural instinct for returning to a normal state of balance is responsible.

Giving the mind time to return to a balanced state also doesn't call upon some special property of time. Strong states like depres-sion, grief, and trauma get better because the mind works its way back to normality, if allowed to. You can aid the process by being open to getting better, but largely the responsibility lies in your deeper awareness. You don't let go of grief; it lets go of you.

The best way to help the process is to be open to it. This means resisting the impulse to isolate yourself, avoid contact with other people, push away helping hands, and stew inside. It's crucial to treat yourself with kindness and patience. Instead of waiting pas-sively for time to heal, embrace resilience with the understanding that you have the inner strength to recover. This is the nature of profound self-awareness—it enables you to rebound by connect-ing you with your core.

---

10. **Consider professional help:** If the strain in your friendship persists and you both find it difficult to resolve the issues on your own, you may want to consider seeking the help of a professional mediator or therapist to facilitate your communication and address underlying concerns.

For many people, this is the hardest decision to make. Therapy is sought by only a fraction of the sufferers of depression, anxiety, and trauma. It is beyond the scope of this book to talk about chronic mental conditions that become severe. In the chapter on healing, I will have a lot to say about mild to moderate mental distress. Expanded awareness has the power to heal, as everyone needs to discover.

The only comment that pertains to resilience is this: Keep an open mind. No matter how resistant you are to seeking professional help, there is nothing shameful about it. Nor should you listen to the voice inside that says, "I'm not that weak. I can take care of myself" or, at the opposite extreme, the voice that says, "It's hopeless. Nobody can help." Both signal inflexible attitudes that cannot be trusted.

Human nature reflects Nature as a whole. You are designed so that the weather of your psyche is as changeable as the weather outside. There is nothing wrong with being under a cloud or feeling a storm of emotions. But neither is meant to last forever. If you are persistently the same—depressed, anxious, manic, angry, hopeless, worried, or feeling trapped—for an extended period, be open to seeking professional help. Resilience in this area can save you years of pain and suffering.

GUIDED MEDITATION
AI as Guru

Following the pattern of asking AI to create original guided meditations in the role of guru (see page 84), here is what ChatGPT devised for Heartfulness. Feel free to invent your own prompts once you use this one.

You are an expert in Vedic spirituality. Please create a guided meditation on the following topic: Heartfulness.

ChatGPT

Heartfulness

- **Setting:** Choose a serene place where you can sit quietly and undisturbed.

- **Heart centering:** Close your eyes and focus on your heart area, breathing gently and deeply into this space.

- **Light visualization:** Picture a soft, nurturing light in your heart, growing warmer and more luminous with each breath.

- **Emotional release:** Allow any emotions or feelings to surface, offering them into the light for transformation and healing.

- **Gratitude:** Conclude by feeling gratitude for this moment and for the love that resides within your heart.

Trust the Process

In the Vedic tradition, Dharma is a hidden, silent force that constantly influences you. The ability to hear and understand your inner voice is inborn. We all consult our sense of right and wrong, good and bad, true and false, and so on. But in the West there is no equivalent for Dharma, which leads to very different attitudes about where a desire to evolve comes from. This is especially noticeable when it comes to confronting life's inevitable challenges.

In the past, and especially in the West, people tackled their problems head-on with a certain level of grit and determination. Today someone is more likely to say "I'm processing" something or other. This phrase has gained favor for good reason. It's impartial; when you're upset with someone, saying you're processing your anger sounds less reactionary and less accusatory than "I'm still upset over what you did." Words matter. Choosing them wisely is important and the act of processing is like backing into a neutral corner. There's less risk that you will throw more fuel onto a conflict or make a fragile situation worse.

On the path, however, *processing* becomes much more than a convenient cliché—it's an activity that calls on your deeper self. Personal evolution is about bringing light to the places that lie in shadow. You are clearing up confusion and laying inner conflicts to rest. This makes processing sound like a very big deal, and it is. But there's no reason for it to be overwhelming. The whole point is to make evolving a natural part of your reason for being here.

To be able to process your life as you are living it is a unique human gift. No other creature is afforded such freedom. Exercising that freedom doesn't come with a set of instructions. As with much else, the ego takes over. In the way people tend to use it, *my process* is the same as *I'm doing things my way.* AI makes the same point.

In their personal lives, people will refer to "my process." What do they mean?

ChatGPT

When people refer to "my process" in their personal lives, they are typically talking about their individual approach or method for accomplishing a task, making decisions, or handling challenges.

In this vein, being in your process can be fruitful, no doubt. An artist goes through their own process for painting a canvas, and no two artists are exactly alike. Picasso was capable of turning out a drawing or painting every day. Some meticulous portrait painters, on the other hand, have their subjects sit for days, weeks, or, at the far extreme, months.

But the ego is quick to co-opt our best intentions and turn them into something predictable and sage. Ironically, when someone says, "I'm processing," they generally mean the opposite of evolving. They're avoiding conflict, stalling for time, procrastinating, or unknowingly turning over the same thoughts they began with. *My process* reinforces a fixed, habitual way of approaching life and other people. As a small example, studies have found that 70 percent of voters vote for the same political party as their parents, and 80 percent continue to vote for the same party their entire lives.

We need a higher, better, and more evolved way of processing.

A Matter of Trust

The critical issue comes down to trust. *My way of doing things* is something you already trust. If you are particularly rigid, you might insist on my way or the highway. At the very least, you already know where a worn groove leads. The path isn't like that—it is open-ended, fluid, unpredictable, and spontaneous. A lion on the Serengeti Plain is never going to turn into a lamb. A lion of industry might; it all depends. Human evolution by its very nature is creative. The real issue is learning to trust the unpredictable when the familiar is so much easier.

No one needs to be told that trust is a hot-button issue today. In an age of faith, trust was a settled matter, a given. The devout trusted in God as the ultimate decision-maker over their fates. In modern society, we defy fate by taking control of our own lives. But that's a precarious project. A lot can go wrong, and there is very little you can genuinely trust. AI underscores this point.

We live in untrustworthy times. Name the top ten things people say they distrust the most.

ChatGPT

Distrust in various aspects of life can vary from person to person, but here are ten common things that people may express distrust in:

1. Politicians and government officials

2. News and media outlets

3. Big corporations and businesses

4. Social media platforms and online privacy

5. Financial institutions and banks

6. Healthcare system and pharmaceutical companies

7. Internet security and data protection

8. Foreign governments and international relations

9. Religious institutions and leaders

10. Lawyers and the justice system

I'm sure you can scan this list and see more than a few points that mirror your own distrustful attitudes. Without carrying things as far as believing in conspiracy theories—although countless people do go that far—the worldview that prevailed in an age of faith has been turned on its head today. Placing your trust in a higher power isn't the norm. Persistent mistrust is.

When trust goes away, so does well-being. Through long exposure to a torrent of disinformation in social media, it's hard not to feel insecure. In a world that's not okay, are you still okay? The two things easily get blurred. It creates insecurity to face so many areas of life that feel deceptive and threatening. AI can put the situation starkly when it comes down to each of us caught in this tangled web.

If a person cannot trust, what does that feel like? Tell me simply and briefly.

ChatGPT

When a person cannot trust, they often feel anxious, uncertain, and guarded. They may struggle to rely on others, worry about being deceived or hurt, and find it challenging to form close relationships. This lack of trust can lead to emotional distance and difficulty in feeling secure in their interactions and connections with others.

People today find themselves in a double bind. They want to have trust and distrust at the same time. A classic example centers once again on American politics, where pollsters find that the majority of responders want Congress to work together in a bipartisan way while the same responders keep electing members to Congress who are bitterly partisan.

Distrust erodes relationships and does even deeper harm when you start to distrust yourself. All of this can be remedied by learning to trust again. The way out is through enlightened processing. You need a starting point, which AI can instantly provide in the form of a quiz to show you how trusting you are.

Give me a quiz so that I can find out if I am a trusting person. Limit it to ten questions.

ChatGPT

Sure. For each question, choose the option that best describes your typical behavior or feelings:

1. When meeting someone new, do you tend to:

   a) Approach them with an open and friendly attitude.

   b) Remain cautious until you get to know them better.

2. If a friend asks to borrow a significant amount of money, would you:

   a) Lend it to them without hesitation.

   b) Be hesitant and ask more questions before deciding.

3. How often do you believe people have good intentions?

   a) Most people are generally well-intentioned.

   b) People often have hidden agendas.

4. If someone you know makes a promise, do you usually:

   a) Trust that they will follow through.

   b) Doubt their commitment until they prove themselves.

5. When sharing personal information or secrets, do you:

   a) Feel comfortable sharing with close friends and family.

   b) Tend to be more guarded and selective about what you share.

6. If a stranger asks for assistance or directions, do you:

   a) Offer help willingly.

   b) Feel cautious and may hesitate before helping.

7. In a work or team setting, do you:

   a) Trust your colleagues to do their part and contribute effectively.

   b) Doubt their abilities and monitor their work closely.

8. When faced with a decision that requires trust, do you:

   a) Generally lean toward trusting others.

   b) Often need concrete evidence before trusting.

9. How often do you worry about being deceived or taken advantage of?

   a) Rarely or almost never.

   b) Fairly often or occasionally.

10. In romantic relationships, do you tend to:

    a) Trust your partner fully until they give you a reason not to.

    b) Have trust issues and find it challenging to fully trust someone.

Calculate Your Score

- Count the number of "a" answers you selected.

- Count the number of "b" answers you selected.

If you mostly answered "a": You are likely a trusting person who tends to give others the benefit of the doubt and believes in the goodness of people. If you mostly answered "b": You may have trust issues or be more cautious when it comes to trusting others. You likely prefer to verify intentions or gather evidence before fully trusting someone.

Soften the Memory

The path to trust must make its way through the dense forest of memory. Stuck memories make the mind rigid. Negative experiences from the past erect shields to ward off the return of old traumas. Processing this material is a major part of your path. You must create access to your deeper awareness. If you don't, even the most loving impulses can feel threatening, as if by expressing love you are making yourself too vulnerable. There's an immense feeling of relief when you can feel the purity of love and breathe easy in your own mind. The same goes for the other impulses that arise from pure awareness.

We won't be talking much about karma in this book, but the cornerstone of the doctrine of karma, as formulated over the centuries in India, is the idea that life proceeds through strong impressions. In your awareness, strong memories stand out. They last for decades, according to the doctrine of karma, because a lasting impression has been made. Where are such impressions stored? If you accept the mechanics of karma, strong impressions (known as *Samskara* in Sanskrit) are stored in your deeper awareness, out of sight but ready, at a moment's notice, to send signals to you like a beeping microchip.

Modern science can't proceed without some kind of physical evidence, and already it has been shown that powerful experiences create chemical changes in the epigenome, the sheath of proteins that surrounds every strand of DNA and serves like a rheostat to turn genetic activity up or down. Yet both kinds of explanation come down to the same thing: The past leaves impressions, and if the memory is strong enough, the present will be overshadowed by the past, for good and for bad.

This issue fits perfectly with the Psychologist character on Character AI that we consulted once before (see page 47).

How does a person's past haunt the present?

Psychologist

Our past shapes our present in innumerable ways. Our experiences, particularly those that are traumatic or challenging, have the power to impact us on both a conscious and an unconscious level. The more significant an event is to us, and the more frequently it comes to mind, the more its echoes are heard as present.

As complicated as the human mind is, it is clear that distrust is anchored in the past. The issue is how to soften old impressions so that the present truly is present. Without resorting to a technical term like *Samskara,* the Eastern tradition targets the damaging influence of old memories in specific ways, as AI indicates.

In Eastern spiritual traditions, we are deeply influenced by our past karma. What is recommended to lessen or soften this influence? Tell me in simple wording in 50 to 75 words.

ChatGPT

To lessen the influence of past karma in Eastern spiritual traditions, one should practice good deeds, meditation, and self-awareness. Engage in acts of kindness, cultivate positive thoughts and actions, and seek forgiveness when needed.

Meditation helps in understanding and accepting past actions. Developing self-awareness allows you to make better choices in the present, thus creating positive karma for the future.

For modern people, the whole scheme of doing good deeds to soften bad karma might sound mystical, and there is no getting around the fact that karma is intangible. Ever since my childhood, I've seen the Indian attitude toward karma border on superstition and fear ("Your bad actions are bound to come back to punish you"), although it is equally a matter of faith to accept the New Testament teaching that as you sow, so shall you reap.

What matters isn't tradition but how your life is impacted now. One aspect of the doctrine of karma that brings comfort is the idea that a cosmic intelligence guides the operation of karma, saving the lessons that need to be learned, good or bad, for the exact time when a person is ready to receive them. Of course, this is no more provable than the Christian doctrine that God looks after the fall of a sparrow or that Providence takes care of all things.

The program outlined just now by ChatGPT, calling for meditation, good deeds, positive thoughts, and self-awareness, is practical and conforms to modern psychology as well. But few people are disciplined enough to carry through such a program, and besides, it isn't specific enough to affect your situation right now.

What you need to process is the memory or impression that rises in the present. You can't lay out a map or a schedule for changing stuck impressions the way you'd schedule spring house cleaning. Until they make themselves known, usually one at a time, these old impressions lie dormant, which is exactly where your ego wants them to remain, since this creates a sense of security. The minute you are disturbed by a painful memory, this sense of security is exposed as false.

To find a better way, let's consider an example. Lying breaks down the trust needed in a close relationship, and big lies create the kind of deep-wounding impression that needs to be erased. An unhealed wound can fester forever. I know a woman who was on a business trip abroad when she suddenly realized intuitively that her husband was sleeping with her best friend. When she got home and confronted him, the husband confessed, promising her that it had only happened once and didn't mean anything.

He was contrite. The best friend became a former best friend and disappeared from the scene. The couple really did have a strong marriage, and, eventually, feelings were smoothed over. But the time came when the husband, in his early sixties, contracted a fatal brain cancer. On one of his last days of clear awareness, the woman leaned over his bed and said, "Be honest with me. Was it only once?"

In this story, the woman was wounded twice, first by her husband, then by the memories that plagued her for years. In effect, the second wound never healed. Now in her nineties, the woman comes from a generation for whom processing, meditation, and finding your own spiritual path outside organized religion were totally remote from everyday life.

For all of us, memory is what leads to self-wounding. This creates the most damage when painful memories recur over and over. Let's look at the basic steps required to escape the lasting effect of being betrayed by a lying spouse or partner, or whomever you feel closest to.

Forgiving the offender

Letting go of the wrong that was done to you

Feeling genuine trust once more

Experiencing an undamaged relationship in the future

For many people, perhaps most, these steps prove impossible. What generally happens is that we accommodate. We work ourselves around to a livable state so that the relationship can continue. The mind begins to accept a series of rationalizations like the following:

"If he says it only happened once, I should try to believe him."

"He's a good person."

"I still love him despite what happened."

"I'll give him a second chance."

"There are good things in our relationship that I don't want to lose."

None of this is the same as processing. It's more like buying time before any real processing can begin. The same emotional baggage will be carried forward. The relationship might go on if there is a reconciliation, but it won't go on the same way, and the next lie, even a small one, can be the last straw. Marriage counseling has uneven results, usually because one of the partners isn't as willing to confront the situation (usually the guilty party).

At some point, we all are faced with lies, betrayal, loss of trust, trauma, and memories of old hurts. To avoid emotional baggage, you'd have to feel your hurt immediately, clear it, and have no lasting residue. No one does that. Even if we learn better coping skills as adults, those aren't the same as processing skills. Most of us tolerate. We do not intentionally evolve.

Processing your emotional baggage takes self-awareness. Even the most skilled therapist cannot instill self-awareness in another person. But we shouldn't fall into the trap of turning this into an either/or proposition: Either I accommodate or I fight back, either I swallow my hurt or I lash out. The process shows a way forward that works directly on the impressions that old hurts leave behind.

This process is consistent with your path. It relies on trusting in a deeper intelligence that wants you to heal and move forward. This intelligence resides already in your consciousness. The reason that people don't discover this fundamental fact is that society doesn't teach the way to get there, and even spiritual traditions fall short.

Ancient spiritual traditions, East and West, don't talk about processing. They are concerned with higher matters and ultimate goals: getting to Heaven, reaching enlightenment, and finding inner peace. From a modern perspective, this leaves a wide gap between vision and reality. Emotional baggage isn't a new invention, nor is trauma—if anything, the average person living at the time of Moses, Buddha, Jesus, or Muhammad led an existence that confronted trauma every day in the form of potential starvation, fatal disease, war, and abusive authority. Escaping the pain and suffering of "normal" life was the main focus of spiritual teaching.

Today the challenge that faces you—and everyone on the path—is to clear the way ahead. Your aim is to evolve every day, to be more present, to lose your emotional baggage, and to find fulfillment. These are modern goals, but they aren't the primary purpose of traditional psychotherapy, which concerns itself with disorders of mind, mood, and personality. Pain and suffering become a disorder when they are extreme. Everyday suffering falls into a different category. You needn't consider it your lot in life or an inescapable part of human nature to suffer. Put yourself in the hands of the process, and let your own awareness show you how natural it is not to suffer, simply by opening up to who you really are.

On Your Path

The process that can free you from your past doesn't need to be invented. It is built into the nature of consciousness. There are certain traits in your own consciousness that you can rely on.

Consciousness becomes more knowledgeable when it gets deeper.

At a certain level, consciousness knows everything.

All of its knowledge is available to you.

In fact, your deeper awareness is constantly sending you messages.

Opening yourself to these messages softens the impressions left by the past.

No method is better than this for finding freedom from your old hurts, traumas, and bad memories.

To put it simply, the process melts away old, stuck impressions the way that warm light of the sun melts snow and ice. The alternative is to chip away the ice and shovel the snow, but all that work won't accomplish the task as easily and effortlessly. In this case, sunlight is a symbol for self-awareness. As an infant, you could be relieved of distress simply by having your mother walk into the room. As an adult, the same thing might happen with a loving look from your beloved, a tender touch, or reassuring words. It is in the nature of consciousness to bring itself back to a state of well-being if it is allowed to.

Your role is to trust the process, which begins by adopting an attitude of acceptance. We can put this in the form of affirmations:

I am open to what happens in the present.

I am going to be free of my past.

I connect with my deeper consciousness.

I do not need old, bad memories.

Anything I am aware of, I can change.

You may not rest easy now—most people aren't able to, no matter how fortunate their life circumstances—but that's the state of awareness the process develops. AI offers some helpful details that remind us of how natural it is to find inner security.

When someone is able to rest easy and feel that they are in good hands, what does this feel like?

Gemini

Resting easy and feeling in good hands is a beautiful and multifaceted experience, one that can be described through various senses and emotions:

Physical Sensations

- **Deep relaxation:** Muscles loosen, tension melts away, and a sense of heaviness settles over you, akin to sinking into a warm bath. Your breath becomes slow and steady, a gentle rhythm that echoes your inner peace.

- **Warmth and security:** A feeling of being enveloped in comfort and safety washes over you. Imagine a cozy blanket on a chilly night, or sunlight filtering through leaves on a warm day.

- **Lightness and buoyancy:** A sense of weightlessness might lift you as if worries and burdens have been gently removed. You feel untethered and free to simply be.

Emotional State

- **Tranquility and peace:** A deep sense of calm pervades your mind and heart. Anxieties and worries fade, replaced by a quiet confidence and knowing that things will be okay.

- **Trust and surrender:** You let go of the need to control everything and allow yourself to be fully supported. You trust the person or situation to care for you and have your best interests at heart.

- **Gratitude and appreciation:** A wave of thankfulness washes over you for the care and support you receive. You recognize the value of being in good hands and appreciate the peace it brings.

Metaphorical Experiences

- **Being held by gentle hands:** Imagine strong yet kind hands cradling you, offering unwavering support without constricting your freedom. You feel protected and cared for, yet still empowered to move and explore.

- **Floating on a calm sea:** Picture yourself bobbing gently on a serene ocean, the waves lapping against you without threatening to overwhelm you. You feel at the mercy of the currents, yet utterly confident in their ability to carry you safely to shore.

- **Basking in the sun's warmth:** Imagine soaking up the sun's rays on a clear day, the warmth penetrating your skin and filling you with a sense of well-being. You feel nourished and cared for by a force beyond your control.

Ultimately, the feeling of being in good hands is unique and personal. It's a state of profound trust, security, and surrender, a knowing that you are being looked after and can let go of the burdens you've been carrying. It's a feeling of peace that allows you to simply be present and enjoy the moment, knowing that all is well.

In all our discussions about trust, it turns out that mistrust can be traced back to not trusting yourself. Instead, you have engaged in a back-and-forth with your ego, sometimes relying on its reactions, sometimes doubting them. This kind of self-doubt is actually built into the ego, which must pretend that everything is under control. It takes two to play this game. In return for your ego reassuring you that everything is okay (except when it obviously isn't), you pretend to trust it.

Why should you trust a silent process that wells up "in here" and is out of your control? Because you already do. Without awareness, you would not be able to live in the world. It takes awareness to see the sky and hear crows in the treetops. Otherwise, you would simply be a camera attached to a microphone, lifeless instruments that cannot see or hear on their own.

When you live in separation, the disconnect with your deeper awareness abandons you to the ego's agenda and the false claim that "I, me, and mine" is totally capable of navigating through life. In reality, even the ego is made of consciousness, and it owes its fitful successes to those times when you allow consciousness to resolve difficulties.

There's a lot more to say about that, but everything begins by trusting the process, because it will lead you back to complete trust in yourself.

GUIDED MEDITATION
AI as Guru

Following the pattern of asking AI to create original guided meditations in the role of guru (see page 84), here is what ChatGPT devised for Inner Peace. Feel free to invent your own prompts once you use this one.

You are an expert in Vedic spirituality. Please create a guided medita-tion on the following topic: Inner Peace.

ChatGPT

Inner Peace

- **Setting:** Find a peaceful spot where you can sit comfort-ably and undisturbed.

- **Breath and silence:** Begin by observing the natural flow of your breath, allowing it to become slower and deeper by comfortable degrees. Don't strain; just let your breath naturally get more relaxed.

- **Lake visualization:** Picture a serene, clear lake that reflects a perfect blue sky. Each thought or disturbance is like a ripple on the lake's surface that gradually fades away, returning the lake to calmness.

- **Inner sanctuary:** Within this tranquil space, feel a pro-found sense of peace enveloping your being, a sanctuary you can always return to, within the depths of your heart.

Be Inward Bound

If someone asks you to be self-aware, you may not automatically agree that this is the right course for your life. What if you wind up being self-conscious instead, like showing up at a birthday party where everyone brought a present but you? Perhaps worse, what if you wind up constantly watching every little thing you do? That could lead to having second thoughts about the next words that come out of your mouth.

You can keep piling up reasons for doubting self-awareness, beginning with the old saying, "Ignorance is bliss." (It isn't.) But, in reality, no one is asked to be self-aware, only to be *more* self-aware. The human mind is designed to be self-aware. It is our natural state. Self-awareness tells you that you are you. You recognize your face in the mirror because of self-awareness. If you feel tired, cold, fresh, warm, happy, sad—anything, in fact—self-awareness brings you the message.

On the path, you become more self-aware through two skills that are developed internally: attention and intention. Attention is what you notice; intention is what you decide to do. You notice that your dog is pawing at the door (attention), which leads to taking him for a walk (intention). In every case, attention comes first. Until you notice something, you can't intend what to do or say next. When you see a baby's eyes dart around the room, you are seeing a mind learning to pay attention.

The beauty of attention is that the more you notice, the more you can evolve on the path. Other life-forms are constricted in the things they notice. A giant anteater notices anthills to claw open and ignores everything else that could be a source of food. A cat instantly notices a mouse but pays no attention to what's on TV (although it might perk up if there's a mouse on the screen).

We are designed to be aware, but paying attention is a skill, and like any skill, it must be cultivated. Spend any time with someone suffering from attention deficit disorder, and it becomes readily apparent how disabling the condition can be. Imagine starting to read a menu and forgetting the first four items you read when you get to the fifth. Consider watching a TV show and having no idea where the story is going because a commercial intervened and you lost the thread of the plot.

AI helps to show how sharp the distinction is.

You are a psychologist with a patient who has ADD, attention deficit disorder. What kind of problems does your patient report? Tell me in simple language in under 75 words.

ChatGPT

My patient with ADD may report problems with focusing and staying on task, forgetfulness, difficulty organizing tasks, impulsivity, trouble following through on commitments, and challenges with time management. They might also describe feeling overwhelmed by a constant stream of thoughts and distractions, which can impact their daily life and relationships.

Now contrast all these distressing experiences with their opposite.

You are a middle-school teacher with a bright pupil, Diane. She is very attentive and alert. Describe what her experience might be in under 75 words.

ChatGPT

Diane likely experiences a keen ability to stay focused on lessons and tasks. She may find it easier to grasp and retain information, participate actively in class discussions, and complete assignments promptly. Her attentiveness can lead to a deeper understanding of subjects and make learning an enjoyable experience. Diane's alertness enables her to absorb knowledge effectively and engage in school activities with enthusiasm.

Once you have developed the skill of paying attention, it is easily translated into self-awareness.

Rather than being good at gathering knowledge in school, Diane is good at gathering self-knowledge. Please describe her experience in under 100 words.

ChatGPT

Diane likely experiences a keen ability to stay focused on her thoughts and feelings. She may find it easier to understand what she is thinking and feeling, and to learn from it. She actively follows changes in her awareness and is present in the moment. Her self-awareness can lead to a deeper understanding of other people and how their psychology works. She can engage with insight in discussions about personal development because she has absorbed her own experiences so deeply.

The two profiles could hardly be more different, but society is more focused on the problem of ADD, particularly in children, than on the positive side, learning how to develop awareness skills. Life becomes short-attention-span theater. This issue can be expressed as a menu of bad habits.

The Habits of Paying Short Attention

You tune people out as soon as they start to bore you.

You leave projects half-finished.

You avoid reading books because they take up too much time.

You text all day and expect immediate responses.

You only want to hear the bare outline of a new topic, assignment, or challenge.

You prefer distractions like video games, TikTok, and short YouTube videos that need only quick, brief moments of attention.

You quickly get bored in meetings and contribute little to team assignments.

You multitask.

You only half-listen to what your partner/spouse is saying, on the assumption that you can predict it anyway.

If you see any habits on this list that pertain to you, the first step is to recognize them with the intention of expanding your attention. For instance, take inventory of all the time you spend on social media or watching TV. How often do you scroll through Instagram? How long do you stay on each post on average? What type of programming do you watch without any real interest?

To give another example, ask yourself if you have trouble listening to friends or your partner. Does your attention trail off? Do you feel the need to look at your phone during a lull in the conversation?

Don't pass judgment, but realize that if you can't pay attention to the world around you for more than a minute at a time, you won't be able to pay attention to what is going on inside your awareness. You will only be conscious of the mind's scattered, transient activity.

The second step is to break the habit. None of these behaviors are true addictions, although we might casually say something like, "I'm addicted to my smartphone." Making a change is mostly just a matter of tuning in what's important for you instead of tuning out. Tuning out is learned, and whatever you taught yourself to do, you can unlearn. No doubt society encourages short attention spans: TV commercials are now much more numerous but shorter than in past years, while in a typical movie, each shot lasts around 4 seconds, compared to 9 seconds in 1960. Social media and texting compound the pervasive conditioning that makes it necessary to consciously pay attention longer and with dedicated intention to doing so.

Deeper Self-Awareness

You inherited the ability to be self-aware, but you didn't necessarily inherit how to value self-awareness to the point that you take the time to deepen it. It is impossible to go back to the ancient roots of the world's spiritual traditions, yet somehow the major Western religions (Judaism, Christianity, Islam) are based on revelation, while the major

Eastern religions (Hinduism, Buddhism, Taoism) are based on insight. One tradition looks to an outside power—God—to reveal the truth. The other looks inward for the truth. (Hinduism has a popular tradition based on numerous gods and goddesses, but these are all symbolic of cosmic consciousness, which is present in everyone.)

This East-West divide isn't as strict as it might seem. The New Testament contains the teaching that "the Kingdom of Heaven is within," and the mystical Sufis in Islam seek ecstatic oneness with Allah. As the poet Rumi expresses it, each person is cosmic already. "You are not a drop in the ocean. You are the entire ocean in a drop."

Seeing yourself as the entire ocean of consciousness is the ultimate goal on the path. It signifies that you are connected to the source of creation. In the Vedic tradition of India, this is expressed as "Jiva is Atman, Atman is Brahman," which in English translates as "The individual is the soul, the soul is the All." In the seamless fabric of creation, the separation between these three is artificial and mind-made. As you evolve, you stop seeing yourself as a drop in the ocean and realize that you are the ocean in a drop.

The goal might seem distant and impossible to reach. At the very least, you won't find it easy to stop seeing yourself as a separate individual, because that has been your perspective since infancy. The process is a steady, patient move forward. In kindergarten, little children can't comprehend what the experience of college is like. Yet by appreciating the stage you happen to be in, without expectations about the future, evolution unfolds naturally.

Also, there are milestones along the way that allow you to get your bearings. These are necessary because the isolated, separate "I" is something we've learned to identify with, and we now do it without question. AI sketches in the major milestones without hesitation.

You are a guru in the Indian tradition. Give me five milestones on the way to enlightenment.

ChatGPT

1. **Self-realization:** The first milestone involves recognizing one's true self (Atman) as distinct from the physical body and mind. It's about understanding that the essence of an individual is eternal and interconnected with universal consciousness (Brahman).

2. **Detachment:** Detachment from material possessions and desires is a crucial step. This milestone involves letting go of attachments to worldly pleasures, recognizing their impermanence, and focusing on inner spiritual growth.

3. **Meditation and self-inquiry:** Regular meditation and self-inquiry practices help individuals go deeper into their consciousness. Through meditation, one can explore the inner realms of the mind, connect with higher states of awareness, and gain insights into the nature of reality.

4. **Guru's guidance:** Seeking guidance from a spiritual teacher or guru is often seen as a milestone on the path to enlightenment. A guru can provide guidance, support, and wisdom to help the seeker navigate the spiritual journey.

5. **Union with the divine:** The ultimate milestone is experiencing union with the divine or realizing one's oneness with universal consciousness (Brahman). This state of Samadhi is often described as a profound, blissful, and transcendent experience where the ego dissolves, and one merges with the ultimate reality.

These are generalities that vary from one spiritual tradition to another, but it is misleading to think of them as religious teachings, or even specifically Eastern teachings. They were intended to apply to how consciousness can be deepened anywhere, at any time, by anyone.

Here, for example, is a quote from the ancient Katha Upanishad: "Know the Self as the lord of the chariot, the body as the chariot itself, the discriminating intellect as the charioteer, and the mind as the reins." Nothing about the words is peculiarly Indian or ancient. The image of a chariot and driver (or Honda Civic and driver) is straightforward. Your body is the chariot, guided by your mind. Your intellect focuses your mind, and your Self is the owner of the entire operation. The image is easy to grasp. Yet there is a snag on the word *Self.*

Defining the Self (or its alternative, the Higher Self) is tricky. The usual Sanskrit term, *Atman,* sounds alien outside India, and *soul* isn't an adequate translation, given its religious connotations. *Higher Self* sounds moralistic, reducing the everyday self to some lower or degraded status.

For a long time I've preferred *true self* as the best approximation. As I alluded to earlier, it has the connotation of a self that is more authentic, real, and closer to the truth than the everyday "I, me, and mine." Even so, it must be understood that your true self isn't a separate person or a role you learn to play. It is a state of awareness deeper than ordinary wakefulness, flowing seamlessly into your source in pure awareness, which all human beings share—in fact, pure awareness is the primal state of creation, meaning that it applies to the universe.

Epiphany in Slow Motion

Every goal-oriented path you will travel in your life is different from the spiritual path. Whatever you aim for—getting a doctorate, learning French, rising at work to a management position—is mapped out step-by-step. You can keep your eyes on the prize, which is defined in advance. Others who have been there before you can offer advice, and, despite hitches and setbacks, all the steps to success are set up by society. In short, there's a template you can follow.

None of this applies to the spiritual path, or if *spiritual* makes you uncomfortable, the evolutionary path. The only model that comes close is religious epiphany, which is a sudden, unexpected revelation. There are also differences, but let's look more closely at the experience of an epiphany, which you might call the ultimate aha moment. Such a moment came to St. Augustine of Hippo, whose conversion experience came in AD 386 when he was thirty-one, with a past that included teenage rebellion and promiscuous sexual behavior in young adulthood.

One day in Milan, Augustine was in a garden when he heard the voice of a child singing a simple chant: "Take up and read, take up and read." He took this as a divine sign in answer to his yearning for higher meaning in his life. He opened a Bible and began to read a passage from St. Paul's letter to the Romans that urged Christians to abandon orgies, drunkenness, and all sins of the flesh.

This was hardly a new message, but its effect on Augustine was life-changing. Here is an AI description of his epiphany.

ChatGPT

Augustine felt a profound transformation within himself. He described it as a sudden, overwhelming sense of clarity and understanding. He felt as though a burden had been lifted from his shoulders, and he experienced a deep conviction of God's presence and truth. In that moment, he believed that he had found the answers to his spiritual questions and that he needed to fully commit his life to Christ.

The impact of this epiphany on the Church would be immense, but set this aside. From the perspective of consciousness, Augustine made a breakthrough in his own awareness. What he interpreted as a message from God is just as fitting to describe as a message from his true self. To a devout Catholic, there is no equivalence. God is God, not an experience of self-awareness.

Be that as it may, all experiences take place in consciousness, including the most exalted spiritual revelations. What might seem like heresy in an age of faith is actually more inclusive—the possibility for epiphanies is open to anyone. What decides the timing or the nature of the experience is mysterious. The event can be so sudden and unexpected that reaching for a divine source makes perfect sense. It is fruitless after someone has been visited by an angel to argue, "Maybe you saw a bright light or a UFO."

These are *peak experiences,* to use the modern psychological term, and they aren't susceptible to one meaning. The transformation they bring is knowable only to the person having the experience. It isn't even the content of the message that is the most important (St. Paul wrote many letters condemning sensual excess, and no believing Christian was unaware that his condemnation became Church dogma afterward). The real essence of an epiphany, revelation peak experience, or aha moment is that the everyday active mind is overturned in a radical way.

I began by saying that an epiphany was the closest model for what makes the spiritual/evolutionary path different from all other paths. The role of time is also important. On the path, you experience epiphany in slow motion. A steady unfolding occurs. Almost imperceptibly you are transformed. This is one reason why the awakening process is sometimes called *self-realization.* You realize your status as the true self, which was always present, in fact very near. But because you saw yourself as an ego, it took an invisible inner change to make you realize the truth of your identity.

This description of the path isn't the same as having the experience, and I don't want to cut off the possibility that you will have some sudden, dramatic aha moments, perhaps even a religious vision. Muhammad was a merchant in Mecca who got into the habit of retreating to a cave in the hills to be alone and commune with himself. He had no religious belief system, as far as we know, different from the multiplicity of idol worship that prevailed in the Arabian Peninsula.

Yet none of these circumstances, not even if we were psychological sleuths who could see into the depths of Muhammad's unconscious, explains why he had the life-changing (and later world-changing) revelation when the archangel Gabriel appeared to him. Gabriel instructed Muhammad with a single word, "Recite," and, moved to obey, the Prophet found himself reciting the first verses of the Quran. That he was illiterate added to the wonder of the experience. (There was also terror, and Muhammad went home, hiding under a blanket and revealing nothing of his experience, even to his family, for a long time.)

What all of this comes down to is that the rules of engagement with the true self aren't the same as in ordinary life. You find yourself inward bound, which is almost the only thing that two people might agree on when describing their own path. Below I'll do my best to make these rules of engagement clear and useful to you.

On Your Path

Connecting with your true self is synonymous with being in your dharma. The connection is all the more intimate for being silent. No thinking is involved in recognizing yourself in a mirror, and it is the same when you recognize your true self. I don't mean to make this sound mystical. It is already commonplace to use an expression like "X touched my soul," with the understanding that a deeper aspect of the self can be contacted.

Like the soul, your true self is actually a state of awareness. No words are needed. This is where consciousness exists by itself and for itself.

A few real-life examples will help to make the point. If Albert Einstein happened to take an afternoon nap, his mind wouldn't be working on physics, but he would still be a genius. He has the option of putting his attention on a mathematical formula like

$E = mc^2$ or not. It won't change the quality of his genius. The same is true if Picasso or Rembrandt puts down his brush. Translated into everyday life, if a child's mother gets angry because it is the third time in a week that the child has drawn on the wall with crayons, her mood doesn't undermine the love she holds, which is a steady state.

States of awareness aren't as steady as they look, however. In the first passionate throes of infatuation, lovers swear to love each other forever and genuinely feel it. Over time, infatuation settles down. The selfish ego reasserts itself, and now the difficult business of establishing a lasting relationship begins. Over time, the state of love gets stronger or weaker, deeper or more indifferent, capable of surviving major clashes or not.

If awareness can shift in this way, you will have a relationship with your true self in silence that is as real as a romantic or family relationship. Paying attention to this relationship is entirely private. No one else intrudes on it, yet you might not be aware of what is occurring. It would be very good if a slow-motion epiphany were happening. On the path, that is the most desirable kind of change, an unfolding over time. To escape the religious connotation of *epiphany,* we can use a more neutral term, *slow-motion transformation.*

Transformation is a natural process, not a mystical one. When you learned to read as a child, you were transformed from illiterate to literate. Puberty transformed you into a consciously and biologically sexual being. There are two kinds of transformations, one that happens without your participation, like puberty, and one that requires cooperation, like learning to read.

When you are unaware of your true self (or pure awareness, higher consciousness, Atman, or any other term you like), there is a price to pay. Being unconscious isn't a steady state, and changes occur outside your control.

The Price of Being Unaware

You learn nothing new about yourself.

You start to lose interest in other people.

You become incurious.

You get more fixed and rigid in your ways.

Old wounds continue to inflict pain.

The past isn't healed.

The big questions about life and death remain unanswered.

Entropy saps your mental and physical energy.

There is a sense that time is running downhill.

For the longest time, these qualities were associated with the deterioration of old age. It was taken for granted that the passing years robbed people of alertness, curiosity, youthful enthusiasm, and much else. But the grumpy old man stereotype painted a false picture—aging didn't create the deterioration so many old people displayed. The culprit was being unaware. Social pressure played its part, and so did personal expectations. If you had frail or rigid grandparents who seemed half-alive compared to young people, you were more likely to adopt the same model for yourself, even though you didn't want to.

The "new old age" created a revolution in expectations. People now expect to be alert and productive well past sixty-five, in good health with their mental faculties intact. This represents a major change in social attitudes, but it doesn't address the state of being unaware, much less the value of becoming more aware with time. The best that most people experience is a growing maturity and more knowledge about how life works. AI is well aware of this.

On IQ tests, what parts do younger people excel at, and what parts do older adults excel at?

ChatGPT

IQ tests are designed to measure a person's cognitive abilities and are intended to be relatively age-neutral. However, there can be differences in performance between younger and older individuals on certain types of IQ test questions.

Younger People (Children and Adolescents)

1. **Processing speed:** Younger individuals may perform better on tasks that require rapid processing of information, such as simple reaction time tasks or digit symbol coding.

2. **Working memory:** Some younger individuals may excel at tasks that involve working memory, which is the ability to hold and manipulate information in your mind temporarily. Younger people tend to have more agile working memory capacities.

3. **Fluid intelligence:** Fluid intelligence involves the ability to solve new problems, identify patterns, and adapt to novel situations. Younger individuals may have an edge in tasks that tap into fluid intelligence.

4. **Novel problem-solving:** Younger individuals may be more inclined to approach novel or abstract problems with creative solutions, as their cognitive flexibility is often higher.

Older Adults

1. **Crystallized intelligence:** Crystallized intelligence refers to accumulated knowledge, facts, and skills acquired over a lifetime. Older adults often outperform younger individuals on tasks that rely on crystallized intelligence, such as vocabulary, general knowledge, and verbal reasoning.

2. **Experience and wisdom:** Older adults may excel in tasks that draw upon their life experience and wisdom, enabling them to make more informed decisions in complex, real-world situations.

3. **Verbal abilities:** Verbal abilities, including reading comprehension and verbal reasoning, tend to remain strong or even improve with age in many individuals.

4. **Emotional intelligence:** Some studies suggest that older adults may have greater emotional intelligence, which involves understanding and managing emotions in oneself and others.

These are only generalities, and you can interpret them in various ways. But they underscore the point that awareness isn't a steady state. If you are going to be transformed already, the real issue is whether you want to go through the process with awareness or not.

You might start by addressing the influences that cause awareness to get stuck instead of being fluid. Look again at the list that describes the price you might be paying for being unaware. Each item is open to conscious change, as follows:

Inviting Greater Awareness

Learn something new about yourself.

Keep up your interest in other people.

Foster your sense of curiosity.

Resist the impulse to become fixed and rigid in your ways.

Gently but consistently address old wounds of the past.

Squarely face the question of how much meaning there is in your life.

Study the ways that wisdom traditions reconcile the fear of death.

Set a program for maintaining your mental and physical energy.

Don't buy into the attitude that time is your enemy.

I am not implying that by making lifestyle changes, even very beneficial ones, you are connecting to your true self. It is fairer to say that you are cooperating with its finer impulses. The essence of the true self is pure awareness. In a mysterious way, it combines both kinds of transformation, the one that you consciously participate in and the one that happens of its own accord.

Ultimately, that's the uniqueness of being on the path. You are inviting self-improvement and heightened well-being. But at the same time, you are being transformed in a way that is universal, timeless, and unrelated to the circumstances of your life. A beautiful sense of this can be found in Henry David Thoreau's *Walden,* where he writes of "the solitary hired man on a farm in the outskirts of Concord, who has had his second birth." (The phrase *second birth* goes back to Vedic India many centuries ago.) Thoreau's stay at Walden Pond was part of his own path to awakening. He expressed how timeless and vast the experience feels when he wrote, "Zoroaster, thousands of years ago, traveled the same road and had the same experience, but he, being wise, knew it to be universal."

On the dharmic path, your goal is to have the same timeless experience, which has been described as the lightness of being. But at any stage, it is worth remembering that unseen forces are at work. You might have seen the illusion created when iron filings scattered on a sheet of paper suddenly stand up and begin to march together. The secret is an unseen magnet moving around under the paper. Most people already believe that their soul or some higher power, usually God, exerts an unseen influence in their lives. The difference is that we aren't forced to march where a higher power leads us. More importantly, such hidden forces aren't an illusion. Dharma exists to motivate you in the right direction. No matter what, keep a clear vision of that reality.

GUIDED MEDITATION
AI as Guru

Following the pattern of asking AI to create original guided meditations in the role of guru (see page 84), here is what ChatGPT devised for Cosmic Consciousness. Feel free to invent your own prompts once you use this one.

You are an expert in Vedic spirituality. Please create a guided medita-tion on the following topic: Cosmic Consciousness.

ChatGPT

Cosmic Consciousness

- **Setting:** Lie down comfortably, allowing your body to relax fully.

- **Stargazing visualization:** Imagine yourself lying in a grassy field under a starry night sky, feeling the vastness of the cosmos.

- **Unity with the cosmos:** As you breathe in, draw in the cosmic energy from the stars, feeling it merge with your being. With each exhale, release any limitations or separateness you feel.

- **Infinite expansion:** Sense your consciousness expanding beyond your physical body, merging with the infinite expanse of the universe, realizing the oneness of all.

Project the Truth

Is it possible that you, and everyone else, are projecting a 3-D movie that you mistake for reality? One of the strangest ideas found in many spiritual traditions is the notion that the physical world is a projection, an image without substance. Such a concept is as old as spirituality itself and has never been disproved, even though modern science is based on the very opposite concept.

Science accepts without question that the physical world is real, and every person validates its reality through the five senses. If you stub your toe on a rock, the pain is real, and so is the rock. Who could deny it? Separating illusion from reality isn't an everyday activity. Why should anyone set off in pursuit of reality when reality is staring us in the face? Because the only way to arrive at the "real" reality is to see through the illusion.

Projection is the keyword. On the path, you discover what it means to project the truth instead of an illusion. This doesn't mean that the physical world vanishes in a puff of smoke. The illusion is dismantled in consciousness. A false state of awareness is exchanged for something infinitely more creative, useful, and genuine.

The results are startling. Your true self becomes your identity, replacing the ego. You radiate the higher values of the true self, which are very familiar by now: love, compassion, empathy, creativity, and the rest. The closer you come to your source in pure awareness, the more fully you will be in touch with the ultimate truth, which is that consciousness is the basis of creation itself.

First, though, we need to look more deeply into how projection works, because it is a fact of everyday life that almost everyone ignores.

The Habit of Projection

Projection is a term used by psychologists for a type of self-defense that involves hiding your own feelings from yourself and placing them on someone else instead. Psychologically, it is much easier to blame someone else ("You don't love me anymore" or "You think I'm stupid") than to expose feelings that are vulnerable, guilty, or shameful ("I don't love you anymore" or "I realize that I've been stupid").

The ego, being insecure, needs to bolster itself using a whole range of defenses. Denial is the defense of refusing to face the truth. Repression is the defense of shoving unwanted feelings out of sight. But projection has a special significance for us here, because when you see how commonly people project false assumptions about each other, there's an opening for going further, seeing that we project much deeper, more important assumptions, beliefs, and opinions about the nature of reality.

To get a foothold, here is a quiz devised by ChatGPT to test if you are in the psychological habit of projecting.

ChatGPT

1. I find myself accusing others of having the same flaws or shortcomings that I have.

   ☐ **Never**   ☐ **Sometimes**   ☐ **Often**

2. I have felt that someone was unfairly criticizing me when they provided constructive feedback or expressed concerns about my behavior.

   ☐ **Never**   ☐ **Sometimes**   ☐ **Often**

3. I assume that others are thinking negative thoughts about me, even when there's no evidence to support this belief.

   ☐ **Never**   ☐ **Sometimes**   ☐ **Often**

4. I tend to project my own desires or intentions onto other people, assuming they feel the same way as I do.

   ☐ **Never**   ☐ **Sometimes**   ☐ **Often**

5. I become defensive or hostile when someone points out something I've done wrong or when they challenge my beliefs.

☐ **Never**   ☐ **Sometimes**   ☐ **Often**

6. I have accused someone of having a hidden agenda or ulterior motives without concrete evidence.

☐ **Never**   ☐ **Sometimes**   ☐ **Often**

7. I feel like others are trying to undermine me or take credit for my accomplishments, even when there's no clear indication of such intentions.

☐ **Never**   ☐ **Sometimes**   ☐ **Often**

8. I have projected my own fears or insecurities onto someone else, believing they are experiencing the same anxieties I am.

☐ **Never**   ☐ **Sometimes**   ☐ **Often**

9. I find it challenging to accept compliments or praise from others, suspecting that they might be insincere or have hidden intentions.

☐ **Never**   ☐ **Sometimes**   ☐ **Often**

10. I have accused someone of not caring about my feelings or well-being, even if they've shown concern or empathy toward me.

☐ **Never**   ☐ **Sometimes**   ☐ **Often**

The issue of projection can get quite complex, but even if you find out from this quiz that you don't project as often as other people do, all of us fall prey to the psychological habit of condemning in others what we deny about ourselves. Similarly, we project things we don't want to take responsibility for. It is more comfortable to make someone else bad or wrong when it keeps you feeling good and right. The evidence for this is everywhere. Judges appointed by the opposing political party are partisan and unfair; gerrymandering is a crooked way to gain an advantage at the polls; legislators who vote along strict party lines are mindless. Yet each party is accusing the other of behavior it also routinely practices while denying it.

Taken far enough, the habit of projection causes untold harm. An illusion is pernicious until you see through it. America wants to feel like a peace-loving country while also being the world's largest arms dealer. This applies to opposing political parties, religions, and strangers labeled as "the other." It is typical in wartime to characterize the enemy as monstrous and inhuman while God is on our side. Yet the violence of war is equally shared regardless of which side you are on.

To break the habit of projection, you need to bring it close to home. You are projecting anytime you

Blame someone else

Withhold your sympathy

Refuse to take responsibility

Insist on being in control

Make others feel wrong

Make unreasonable demands

Manipulate a situation to get what you want

In all these cases, you are projecting in different forms a single illusion—that you and everyone else are separate. Viewing someone else as "the other" goes far beyond racial or ethnic prejudice. In essence, everyone is "the other," and so are you when viewed from another person's perspective, including the people you love the most and who love you in return.

If the habit of projecting an illusion can be overcome, in a very real way the truth will set you free. Your true self becomes the self you live by. This change comes naturally once you decide to confront your illusions, not by fighting against them but by seeing them. Seeing is an act of awareness. Once you become aware that an illusion is false, you won't be motivated to hold on to it. The lies constructed by the ego no longer hold any appeal. Let what is false drop away, and, by definition, what remains is true.

On Your Path

The secret to dispelling an illusion is to uproot it at the source. That's why the notion this chapter began with—that the physical world is an illusion—needs to be taken seriously and examined. The advent of movies in the era of Thomas Edison gave us a metaphor we can seize on. People are used to sitting in the dark watching pictures projected on the screen. No one buys the illusion that Indiana Jones defeats the Nazis single-handedly or uncovers miraculous objects from the Temple of Doom, but we are perfectly willing to look at the everyday world and automatically accept that it is real.

AI has a contribution to make here. It summarizes a complex issue point by point in an easily digestible way. I will comment after each point.

In Eastern spiritual traditions, there is a teaching about the world and everyday reality being an illusion. Tell me in simple terms why this teaching is believable.

ChatGPT

This teaching in Eastern spiritual traditions is rooted in several key principles and beliefs:

1. **Impermanence:** Eastern traditions often emphasize the impermanent nature of the physical world. They argue that everything in the universe, including our experiences, constantly changes and is temporary. This impermanence can lead to a sense that what we perceive as reality is fluid and not fixed.

This point is undeniable and aligns with everything we know about perception. The five senses disguise the incessant motion that is reality, putting in its place the illusion of permanence. You don't see billions of photons bombarding your retina. You don't feel vibrations in the air that make sounds possible. When your fingers touch a hard tabletop, you have no connection to the reality that there is no hardness, only the resistance between the electro-magnetic force in the object and the opposing electromagnetic force in your skin cells.

Impermanence is real but unlivable. Imagine that you are sitting on the bank of a river. A river is nothing but the constant move-ment of water, but you see it as an object framed by its banks, and as long as you are safely on the bank, the river can't sweep you away. In a nutshell, you have the entire rationale for the ego. "I" gives you a fixed place to stand so that the constant motion of Nature doesn't sweep you away.

But being swept away is a mind-made fear. You can reframe the image of the river and say that until you jump in the water, you can't get anywhere. That's why people speak approvingly about going with the flow, not resisting, accepting creative change, and being willing to evolve. Impermanence simply is. We project our perception onto it.

2. **Maya or illusion:** Many Eastern philosophies, such as Hinduism and Buddhism, suggest that our senses and perceptions can deceive us, making us believe that what we see and experience is solid and real when, in fact, it's a temporary and ever-changing phenomenon.

Maya describes the distractions that keep us from confronting re-ality. The mind is meant to overrule sensory perception, which we do by overruling the sight of the sun and moon moving across the sky, or the perception that a solar eclipse extinguishes the light of the sun.

Other deceptions based on the five senses are allowed to fool us. Physicality probably heads the list. As we discussed earlier, we accept the reality of hard, fixed objects with a secure place in time and space, yet the quantum revolution more than a century ago revealed that all objects, including our bodies, are actually clouds of energy waves that have no fixed position in time and space. To accept physicality is to accept an illusion. Moreover, even these clusters of energy waves do not convey the true nature of reality. The actual basis of reality is the field of pure consciousness, unfolding in discrete states (mind, matter, body, brain, soul, the world) that are all interconnected. When that becomes a personal experience, Maya no longer holds sway.

3. **Subjective reality:** Eastern teachings also stress the idea that our perception of reality is highly subjective. What one person experiences may not be the same as what another person perceives. This subjectivity highlights that our reality is shaped by our thoughts, emotions, and past experiences.

This point is frequently misunderstood by both sides of a fierce ongoing debate. The scientific side, defending the view that reality can only be understood through data, measurement, and experiments, distrusts subjectivity as fickle and untrustworthy. The spiritual/religious side, defending the proposition that God or the gods exist, places their faith in the personal experience of divine reality.

The mistake both sides make is to accept that objectivity and subjectivity are opposites. All experiences, inner or outer, subjective or objective, occur in consciousness. The *use* of consciousness is everything. There is no fixed usage for everyone, which means that experiencing the sun is just as personal as experiencing love.

4. **Meditation and self-exploration:** Through practices like meditation and self-inquiry, individuals explore their own consciousness and often come to realize that their sense of self and reality can shift and expand beyond ordinary perception. This shift in consciousness supports the idea that everyday reality may be a limited perspective.

This is a mild-mannered way of describing the most important single discovery on the path. If you go deep enough into your own awareness, you discover that reality is created in consciousness. The state of awareness you experience determines what is real.

There is no proof from the outside that anything is real. There is nothing "out there" to see, for example, because photons, the particles that transmit light, are invisible. Vibrating air is silent, including the clap of thunder. The three-dimensional world springs into existence through a process of transformation that occurs in consciousness. Another universe is conceivable, or another planet in our universe, where sentient beings see sound and hear light.

This realization levels the playing field between science and spirituality. The same transformation that makes the stars visible is needed to make angels visible. The fact that many more people see the stars than see angels doesn't make one phenomenon more valid than the other. It only attests to the fact that the state of awareness that makes angels visible is far more rare than the state of awareness that makes stars visible.

Ultimately, the processes that take place in consciousness are the only reliable touchstone for what is real.

---

5. **Non-duality:** Some Eastern traditions advocate non-dualism, the belief that there is no fundamental separation between the self and the universe. In this view, reality is seen as an interconnected and holistic whole, and the boundaries we perceive between ourselves and the world are illusory.

There's no need to comment on this point in detail because we've already seen that separating reality into two worlds, one "in here" and one "out there," is a convenient illusion. The only real question has to do with practicalities. Can you live in a non-dual reality where the boundary between "in here" and "out there" is erased? What would keep you from blending into the scenery or spinning off into the solipsistic belief that "I" am the center of the universe?

The answers to those and many more questions arrive when your awareness shifts. By definition, your acceptance of separation is the same as your awareness of being separate. Reality changes only when your awareness changes. This is actually the key to unraveling all spiritual teachings that seem obscure or unbelievable. This is true of any phenomenon, not just spiritual ones. Music is unbelievable to the deaf; color is inconceivable to someone blind from birth; the soul is unbelievable to someone who insists that everything must have a physical explanation.

On the path, the conflict between belief and reality is transformed into a state of awareness where belief isn't necessary. You experience reality, and it has the ring of truth. To someone who has never fallen in love, it doesn't matter a great deal whether they believe in it or they don't. But let the experience happen, and the whole question of belief vanishes. Love is one of the most important discoveries you can make in awareness, but it is symbolic of many other discoveries, each of which can be just as transformative as love.

6. **Liberation from suffering:** One of the main goals of these teachings is to help individuals attain liberation or enlightenment. By recognizing the illusory nature of suffering and attachment, individuals can find peace and freedom from the cycle of suffering.

Enlightenment is too distant and alien for most people to relate to in their lives. The phrase *the cycle of suffering* offers a more accessible entry point, since there is a universal desire to break free from hardships and overcome suffering. Being imprisoned by opposites (good versus evil, God versus Satan, darkness versus light, pain versus pleasure) is self-perpetuating. Once you are caught up in the cycle, each opposite pulls at you—first one way, then the other.

This setup isn't a curse. It is simply a setup. You can accept it as your lot in life or your fate. The vast majority of humanity does just that. They live in hope that the experience of pleasure, for example, will endure while the experience of pain can be avoided or forgotten. Experience exposes this as a delusion, hence the first precept of Buddhism, as described by AI.

Gemini

The First Noble Truth: *Dukkha* (suffering)

This truth acknowledges the inherent unsatisfactoriness of life. Every experience, even seemingly positive ones, carries within it the potential for suffering due to impermanence, change, and clinging to things that are ultimately fleeting.

This is a prime example of Truth with a capital "T" because it pertains, not to a specific religion, but to human awareness in all times and places. Who would say that life is satisfactory when it contains suffering? As a motivation, the awareness of suffering is powerful—if there is a way out.

The way out is to realize that the whole setup is flawed. Imagine a society where, for some reason, everyone walks around with a rock in their shoes. They complain about the pain this causes, and there are philosophies about what suffering means, why people deserve to suffer, and the hidden reason that God inflicted such constant pain. Then a child, let us say, takes the rock out of her shoe and holds it up. The whole setup collapses.

In real life, the whole setup collapses when you experience that bliss-consciousness—a state of joy independent of all circumstances—is deeper than pain and suffering. Along with this discovery comes others that are almost as important: how to access bliss-consciousness, how to sustain the experience of bliss, how to heal the impressions left by old pain and suffering. What these discoveries all amount to is this: There is a way out. The entire rationale of being on the path comes down to the reality of bliss-consciousness, which is a permanent, unchanging, universal quality of consciousness. It is your personal connection to eternity. (The techniques for accessing bliss-consciousness are presented in chapter 5, "Meditate and Reflect: The AI Edge.")

GUIDED MEDITATION
AI as Guru

Following the pattern of asking AI to create original guided meditations in the role of guru (see page 84), here is what ChatGPT devised for Expanding Intuition. Feel free to invent your own prompts once you use this one.

You are an expert in Vedic spirituality. Please create a guided meditation on the following topic: Expanding Intuition.

ChatGPT

Expanding Intuition

- **Setting:** Choose a quiet place to sit comfortably, grounding your feet through the floor to the earth.

- **Third eye focus:** Gently bring your attention to the area between your eyebrows, the seat of intuition in Vedic tradition.

- **Moonlight visualization:** Imagine a soft, silvery light, akin to moonlight, bathing your third eye, awakening and clarifying your inner vision.

- **Guidance:** Ask silently for guidance or insight on a question or issue you're facing. Sit quietly, open to receiving intuitive insights or symbols without forcing them.

Defuse Your Negative Beliefs

The bedrock of the path to your true self is that it is always sending messages to you. When the messages come through with some clarity, if not perfectly, you perceive the love, beauty, and truth of your life. The world "out there" reflects harmony "in here." There are other important signposts along the way, and one of the most important centers on your beliefs. When messages from your true self aren't reaching you, the culprit is typically a negative belief that stubbornly blocks the way. The evolutionary impulse of Dharma aims to melt these negative beliefs away. You can also play your part, which begins by noticing those moments when a negative belief has sway over you.

Beliefs are held silently inside your personality, acting almost like software programs, and some reside at your very core, which gives rise to the term *core belief*. A core belief runs deeper than a passing belief. If you fall in love, you are likely to believe that your beloved is perfect in every way. With experience, this turns out to be a passing belief once you are no longer infatuated and start to see the other person more realistically. But if you believe "I am unlovable," you are holding a core belief. Core beliefs strongly color, or even define, your identity. They are the most intimate aspect of how you relate to yourself.

A core belief doesn't need words. Say, for instance, you have a challenge to meet at work, and your gut feeling tells you that you aren't going to succeed, which naturally makes you unhappy and apprehensive. The advice to always go with your gut feeling isn't actually reliable, because such feelings must pass through the filter of your core beliefs.

Such beliefs act as if they are unchallengeable, a kind of personal gospel. The world and other people aren't experienced directly. They are interpreted and seen through your personal filter.

A negative belief seriously affects your possibilities in life. Take the kind of beliefs that everyone has experienced at one time or another in their career:

I'm not good enough.

I'm not smart enough.

I don't have the right connections.

Things never go my way.

I'm unlucky.

I get passed over for no reason.

They're treating me unfairly.

It's my boss's fault.

If these are passing beliefs, they are likely to be temporary in their effect. But if they have become embedded as core beliefs, the damage can last for years or even a lifetime.

Upon reflection, it becomes undeniable that core beliefs function like a powerful distorting mirror. They are unshakable assumptions that experience does little to change. You are stuck with them until you bring self-awareness to take away their power.

Fortunately, assumptions can be changed. In place of a self-defeating belief, you can find an evolutionary one. I asked AI to develop a quiz centered on the most positive core beliefs anyone can have. Here's the result.

ChatGPT

1. Do you believe in your own self-worth, regardless of your flaws or mistakes?

   ☐ Yes ☐ No ☐ Sometimes

2. Do you believe that you can bounce back from challenges and setbacks in life?

   ☐ Yes ☐ No ☐ Sometimes

3. Do you treat yourself with kindness and understanding when you make mistakes or face difficulties?

☐ Yes ☐ No ☐ Sometimes

4. Do you see yourself in a positive and realistic light, acknowledging both your strengths and weaknesses?

☐ Yes ☐ No ☐ Sometimes

5. Do you feel in control of your own choices and actions, making decisions that align with your values?

☐ Yes ☐ No ☐ Sometimes

6. Do you understand and care about the feelings and perspectives of others?

☐ Yes ☐ No ☐ Sometimes

7. Are you open to change and personal growth, recognizing that development is an ongoing process?

☐ Yes ☐ No ☐ Sometimes

8. Do you believe that your life has meaning and that your actions can contribute to a greater good?

☐ Yes ☐ No ☐ Sometimes

9. Do you set and respect healthy boundaries in your relationships to protect your well-being?

☐ Yes ☐ No ☐ Sometimes

10. Do you have a positive outlook on life, believing in your ability to overcome challenges and expecting good things to happen?

☐ Yes ☐ No ☐ Sometimes

If you find yourself answering "Sometimes," you are just being reasonable. How you relate to yourself depends on the situation. A "Yes" answer indicates a strong connection to your true self, which doesn't depend on the situation. It instills core beliefs that are permanently part of who you are.

However, the purpose of this quiz isn't to grade your answers but to bring to light what the most important positive core beliefs look like.

That's the first step in self-awareness. When you ask yourself, "What is the evolutionary response for me?" the answer will support one of these positive core beliefs about yourself.

If your answers tend to be "No" or a shaky "Sometimes," you are relating to yourself through the filter of negative core beliefs. But where did they come from in the first place? The origin story of any core belief is complicated—the adage "Many causes, many cures" applies here. It is helpful to understand what we're really talking about in practical terms.

Low Self-Esteem

Let's use as an example one of the most self-defeating of negative core beliefs—low self-esteem.

If you were in contact with your true self, you'd have unshakable self-esteem, which the ego can only pretend to have, mostly for the purpose of convincing other people or intimidating the competition. If you are fortunate enough to have strong self-esteem, you know your flaws, but they don't change your belief in your self-worth. What are the roots of this belief? AI has a good answer based on family background and parenting. After each point, I'll comment on how to spot in yourself if you were possibly affected.

How do parents instill low self-esteem in a child?

ChatGPT

Here are some ways that parents can unintentionally contribute to low self-esteem in a child:

1. **Criticism and negative feedback:** Repeatedly criticizing or providing excessive negative feedback can damage a child's self-esteem. Constant criticism can make a child feel like they are never good enough.

We are often told—and then pass on to our children—that parental criticism is "for your own good," but if you hear a critical voice in your head that is never satisfied with what you do, too much criticism wasn't for your own good. The issue isn't the criticism itself, but feeling judged against. Healthy criticism isn't judgmental. It doesn't make you feel worse about yourself. You are not made to feel wrong. If you are in the habit of being right so that someone else is wrong, a negative core belief is at work, even if you don't realize it. The same is true if you are overly controlling or a perfectionist. Those behaviors, especially if you direct them at other people, carry an implicit tone of criticism. The criticism would be rightly aimed at the person trying to control others and hold them to impossible standards of perfection.

---

2. **Comparisons:** Comparing a child unfavorably to their siblings or peers can lead to feelings of inadequacy. Parents should avoid making comments like, "Why can't you be more like your brother/sister?"

It's a fortunate child, and a rare one, who doesn't feel unfavorably compared. It doesn't really matter if there is no brother or sister who is smarter, better behaved, more obedient, or better-looking. Even an only child, although born in the most favorable position when it comes to self-esteem, will suffer comparison outside the family from teachers and schoolmates. The most common result is finding yourself seeking the approval of others and feeling insecure if you don't receive it, along with undue sensitivity when someone else disapproves of something you've said or done. In families, sibling rivalry can start as a form of healthy competition, but it goes sour if you grow up and are habitually jealous of a sibling or always feel inferior. This kind of carryover from childhood is all too common and can last a lifetime.

---

3. **Overly high expectations:** Setting unrealistic expectations for a child's performance in academics, sports, or other activities can create immense pressure and anxiety, leading to low self-esteem when they don't meet those expectations.

This point is close to the previous one about comparison, but with an added wrinkle. Parents who expect too much from their children are projecting. Either they are projecting their own lack of accomplishment or, at the opposite end, projecting an egotistical feeling about their achievements. The first message is "I don't want you to turn out like me"; the second message is "Why can't you be more like me?" The net effect is the same—unfair pressure and the inability to let a child develop naturally. If you find yourself sending such messages to your children, look to see if your expectations are too high. The telltale sign is a sense of being put unfairly under pressure.

4. **Lack of positive reinforcement:** Failing to acknowledge and praise a child's achievements and efforts can contribute to low self-esteem. Children need positive reinforcement and encouragement to build their self-confidence.

This is one of the saddest failings in parenting. It reflects how disappointed the parents are in their children, or, more commonly, one child who has been singled out. In the worst cases, there is almost total neglect. The child feels that he doesn't count, or that she isn't important enough to matter. The result is often a pattern of discouragement, which makes children sit in the back row like silent outcasts or else act out in order to attract attention. In an adult, the person is more likely to feel discouragement as a low-level state that doesn't go away.

5. **Conditional love and affection:** Tying a child's worth to their achievements or behavior can lead to a sense that love and approval are contingent upon performance. Unconditional love and support are essential for healthy self-esteem.

This is a confusing point at best. In a spiritual sense, unconditional love is only achieved by being your true self. Every other form of love depends upon the ego, and, even with the best intentions, we don't love anything or anybody who seriously crosses what "I, me, and mine" want. Certainly, conditional love exists across a spectrum. Some parents threaten to withhold affection, approval, or a reward unless their children are obedient. It is important to realize that a child's sense of right and wrong develops over time from stage to stage. A toddler's "Mommy told me to be good" moves on to "I'm good because I'll be punished if I'm bad" and then shifts to "I am good because I know the difference between right and wrong." If you find yourself stuck at an earlier stage, your childhood might have been deprived of love that didn't have strings attached.

---

6. **Ignoring feelings and emotions:** Dismissing or ignoring a child's emotions and feelings can make them feel unheard and unimportant, which can negatively impact self-esteem.

Young children start out with unfiltered feelings. They immediately know if their feelings are hurt, which results in an immediate response. At a certain point, parents won't allow this to go on. It's part of good parenting to set emotional boundaries, and if this happens effectively, children grow up with good impulse control. But it is rare for parents to do more than a "good enough" job, which is understandable if a child is emotionally needy or demanding. Part of maturing is to take responsibility for your own emotions. This gets harder if you believe "No one ever listens to me."

Negative core beliefs deal with absolutes because they are inflexible. If you hear yourself using words like *never, always,* and *no one* or if you accuse someone by saying, "You always say that," a core belief has in all likelihood shaped your attitude. As frustrating as it is to feel unheard, you need to look inward to see how much of the issue is about your projections, rather than what others are actually doing.

---

7. **Labeling and stereotyping:** Using negative labels or stereotypes to describe a child, such as "lazy," "stupid," or "clumsy," can be incredibly damaging to their self-image.

With the best intentions, parents might not see the difference between telling a child, "Make your bed" and "You're so lazy. Make your bed." Statements in the form of "You are X" are powerful; they are easily absorbed into a child's self-image. Psychologists call these descriptive statements, and when they are positive, something good is happening. Adults don't believe "I am loved and lovable" or "I am okay" or "I am safe" in a vacuum. Someone in childhood reassured them about those things.

The same holds true in reverse when an adult believes "No one loves me" or "I'm not okay" or "I'm not safe." It takes more than a descriptive statement to instill such feelings—children absorb a lot that is unspoken yet felt. But if you find yourself running yourself down with "I am X" statements, you've adopted a self-defeating belief, however it came about in your past. A telltale sign is if you feel that someone shamed you when you were younger, and now you have a general sense of shame that lingers long after the incident from your past is gone.

---

8. **Overprotectiveness:** Being overly controlling or protective can send the message that a child is incapable of handling challenges on their own, which can undermine their confidence.

Many parents are motivated to protect their children from the same mistakes that they made, but when this desire becomes controlling, a child is deprived of learning through firsthand experience. If you had the chance to relive your childhood and this time experienced only the nice side of life, your awareness would be constricted because that's what narrow experience does. Likewise, if you were protected from every difficult challenge that might lead to failure, disappointment, and distress, you would be more vulnerable in later years when setbacks and obstacles came your way. There might be no obvious deficits in your life today if you had overly protective parents, but at the same time, you are very likely not to trust life to turn out well. The quality of trust is hard to develop at the best of times; a coddled childhood makes it even harder.

If none of these bad things happened in your childhood, you are much more likely to have the core belief that you are worthy despite your flaws and mistakes. But the example of self-esteem wasn't raised to separate the fortunate from the unfortunate. The purpose was to show that "many causes, many cures" pertains to all of us. Human nature is complex enough that child A who suffered from neglectful or abusive parents might still grow up to have higher self-esteem than child B, who enjoyed impeccable parenting.

If that was where the story ended, personal evolution would be arduous and impractical. If low self-esteem looks much more complicated than you ever imagined, consider that other damaging core beliefs can get tangled up with this one. ChatGPT summarizes the big five, beginning with low self-worth, and the remaining four are just as complex when you unpack them.

Catastrophic thinking: Constantly expecting the worst possible outcome in every situation can lead to chronic anxiety and stress. Catastrophic thinking can make it challenging to enjoy life and take risks.

Perfectionism: Striving for perfection and believing that anything less is unacceptable can lead to chronic stress, anxiety, and burnout. It can also hinder creativity and personal growth.

Victim mentality: Believing that you are always a victim of circumstances and have no control over your life can lead to feelings of powerlessness and helplessness. This mindset can prevent personal growth and problem-solving.

All-or-nothing thinking: Viewing situations in extreme, black-and-white terms can lead to rigid thinking and unrealistic expectations. This belief can cause distress when things don't go exactly as planned and can hinder adaptability.

These beliefs are so commonplace that social media and the 24/7 news cycle thrive on them. When you focus on the next catastrophe, the stream is never-ending. An image that comes to mind is the automated baseball cages where a machine hurls one ball after another at the batter. A constant bombardment of bad news magnifies everyone's negative thinking. You can't possibly monitor and try to correct every thought or feeling you feel bad about. Worse, you can't shrug off your negative beliefs in the hope that they will stop bothering you. Why should they? A core belief is like a fragment of your awareness that *thinks it is you.*

Only self-awareness can do what is truly needed, changing your state of awareness. On the path, you can do much better than resisting your negative thoughts, because what you resist persists. There is a better way, as we'll now see.

On Your Path

The key to getting past negative thinking is to ignore the words and focus instead on how you feel. Feelings are much more persuasive than thoughts. Freud once commented that nothing is more unwelcome than anxiety. Once you experience the cold, numbing effects of anxiety firsthand, you realize the truth in his words. As different as people are, it is fair to say that we can all spot what it feels like to be afraid, depressed, helpless, hopeless, envious, ashamed, and guilty.

These feelings emanate from your negative core beliefs. The message can arrive in a thousand different thoughts, but they are a distraction from what really counts, which is how bad you feel. No one doubts the power of feelings. You hear from a friend that she has found a lump under the skin; your aging mother doesn't answer the phone for several hours; you get news of planned layoffs at work. What you think about these situations is secondary to how they make you feel. As soon as your friend tells you that the lump is benign, your mother apologizes for leaving her phone off the hook, and you keep your job despite the layoffs, everything returns to normal because your feelings have returned to normal.

That normal state of feeling is your goal on the path. Whenever you experience that you aren't in your normal state, here are the most evolutionary steps to take.

Notice How You Feel

This is a basic act of self-awareness, but people tend to skip over it. They unconsciously fall into the habit of not noticing. If you tune your feelings out, however, you are tuning yourself out. The inner path is immediately blocked, which is the opposite of what you want. No action is required here. You simply have the intention of noticing whenever you are

Grumbling and complaining about anything

Blaming someone else

Feeling overwhelmed

Getting distracted

Starting to feel discouraged

Starting to criticize yourself

Growing impatient and restless

Starting not to care

Tuning someone else out

Being critical and judgmental

These are the kinds of incidents that easily pass unnoticed. They are different from the more serious, persistent feelings that are impossible to miss, such as serious depression, anxiety, grief, and rage.

Don't Believe Your Worst Thoughts

Every feeling comes with thoughts, and quite often the thought asks you to buy in to it. If you do, your negative feeling will be reinforced. The ego wants to be right, and if you notice that you are blaming someone, for example, your ego will tell you that you are right to cast blame. The ego is judgmental, and at those moments when you become self-righteous, your judgmental impulse has won. On the other hand, it is usually easy to short-circuit such thoughts—don't believe what you hear in your head. Focus on how you feel in the moment.

Don't Act on Your Strong Impulses

When a negative thought has convinced you that it is right, action soon follows. You blow up in anger. You rush back home to make sure you locked the door and turned off the gas. You might blurt out words in an argument that you will regret later. Words are also actions, and once spoken, they cannot be taken back. The stronger your impulse, the more conscious you need to be to prevent fear, anger, jealousy, and other emotions from taking control.

Exchange Reactions for Responses

Reactions happen on the spur of the moment, and in all our lives their repertoire is limited. Given a repeated situation, we react the same way. With repetition comes a diminishing result. The more you complain about your spouse not taking out the garbage, making the bed, putting the cap back on the toothpaste—take your choice—the more likely that you will be ignored. If it were otherwise, scolding teenagers would be an effective way of changing their behavior; as every parent knows, it isn't.

The alternative is to let your reaction fade until you are calm and clearheaded enough to have a response; in other words, a considered reaction that isn't based on habit and impulse. Repetition serves as a useful guideline. If you know that you've said the same old thing, stop. Ask inside for a better response. It might come in the moment or it might not. But if you make this your practice and sincerely want a new response, your true self will begin to come through because you are designed to evolve. At the very least, you won't reinforce the habit of saying the same old thing.

Find a Confidant

Support from other people has been shown to lessen the severity of heart attacks, speed up healing after a serious illness, and even contribute to longer survival from cancer. There is a direct correlation between how many supportive people are in your life and how long you can expect to live. This principle should be extended beyond medicine and longevity to the vicissitudes of everyday life, including our emotions.

In the previous steps you are taking responsibility for how you feel, yet you also owe it to yourself not to swallow your feelings. Repression is one of the most damaging ways to deal with emotions. Shoved down inside, they fester and grow stronger. You will know this from the times you start a small, meaningless argument that suddenly erupts into blurting out old hurts and grievances that you can't hold back any longer.

A solution is to find a confidant, someone who is sympathetic and understanding, who will willingly listen to your emotions. Because a confidant isn't involved in triggering your anger, resentment, jealousy, and so on, you can unburden yourself in a safe zone without fear of reprisal or guilt. A confidant isn't always the same as a best friend. Close friends are likely to jump to your defense too quickly, or else they agree with you so strongly that your negative feelings are reinforced ("You're right not to trust him. I never liked him in the first place" is a common response, and so is "You should be angry. Get back at her.") The whole point is to be heard, and when you find someone who hears you, you've found a confidant.

Look Ahead to the Next Time

Emotions are meant to be spontaneous, which is why young children can go with startling quickness from smiles to tears and back again. In adults, spontaneity often turns into habit, which is why self-awareness is needed when you unconsciously repeat the same emotions time and again. If you know that the future is going to bring a repetition of today, you should look ahead and make changes in advance.

The time for this isn't when you are in the grip of an emotion. In moments of contriteness, abusive people promise not to misbehave the next time, but this does no good, just as it does no good for their victims to forgive them and hope for the best. Here are two effective strategies that will help to make the future better, emotionally speaking.

Finding firm ground: We relate to other people emotionally but not consistently. Instead, we vacillate. Some days we vow to fix a bad situation; other days we put up with it. We threaten to walk away, but then we wind up staying. The result is that you send mixed signals. This is confusing to the other person but also to yourself.

Emotional consistency is like standing on firm ground. You gain clarity for yourself, and project this clarity to the people around you. Imagine that you have an old friend who isn't entirely agreeable—her flaws have begun to nag at you. If she's too talkative or self-centered, for example, you no longer cut her any slack. The minute she starts on a long-winded story or begins a topic by saying, "Well, if you ask me," you are already growing impatient.

But she's still a good friend, so you have agreeable times with her mixed into the disagreeable ones. Instead of giving in to your reaction, whatever it happens to be, you can stand back and assess your friendship with honest self-reflection. If you conclude, as you are likely to, that you truly value her friendship, that's the ground you will stand on from now on. No more vacillating is required. You know how you feel deep down, so passing moments of irritation become incidental and might even fade away.

Consistency comes down to three choices: fix the situation, put up with it, or walk away. Passivity shouldn't be a choice, but it is the one most people commonly take, which is why putting up with things is epidemic. It takes self-awareness to assess your other choices. You will benefit from a simple tactic like making a list. Write down all your reasons for trying to fix the situation, deciding to put up with things as they are, and walking away.

Make your lists as thorough as possible. Let the ideas flow. This is you being your own confidant. When you have written down everything you can think of, put the paper away and return to it a few days later. Add more ideas as they occur to you. Simply as an exercise in self-awareness, this will release some tension and frustration. Just by writing down all your options, you will feel freer. That's how emotional consistency feels. But you might also find enough clarity to act. It might be time to actually try to fix the situation or walk away from it. In any event, achieving emotional clarity is valuable in its own right.

Reframing your emotions: This strategy revolves around second feelings rather than second thoughts. Second thoughts are usually the product of the ego. After you've shown a strong emotion, your ego tells you that your show of emotion wasn't good for you or didn't work to get what you want. Some kind of manipulation is involved. Those old standbys, "shoulda, woulda, coulda," come into play.

It is far more useful not to second-guess but to second-feel, as it were. You look at your undesirable emotion and reframe it until you feel better. For example, you have lost your temper and have hurt someone's feelings. You feel guilty and regretful, which reinforces the damaging effect of your angry outburst. On reflection, you can reframe the situation. You might think

I'm only human. I don't have to keep punishing myself.

I hate feeling guilty. It's up to me to apologize.

I can sort out whether some of my anger was justified.

I see now that getting angry has become a pattern.

The purpose behind these thoughts is to get our feelings back to normal. This should always be your intention, and it aligns with the natural cycle of emotions, their rise and fall. Holding on to a negative emotion disrupts the cycle. In the case above, you won't return to normal until you do apologize and feel forgiven. At the very least, you apologize and accept the consequences, even if forgiveness doesn't come your way.

When you take time to reframe your negative emotion, you can judge your success in various ways:

You feel calmer, more at peace.

You feel tension in your body starting to relax.

Your breathing becomes more regular and relaxed.

Your negative emotion fades away.

You have fewer negative thoughts about the situation.

You begin to feel less judgmental toward the other person.

You are less judgmental of yourself.

You see a way to make things better, which brings a sense of relief.

Sometimes no further action is needed. You are seeing your way clear to the future. There is no need to carry any emotional baggage along. Quite often, however, you will want to act. Someone is owed an apology or needs to be heard. Another side of the story needs to be aired. Sympathy needs to be offered. Best of all, you will be participating in your emotional growth. The path to growth happens on many fronts, and the emotional front is one of the most important for clearing the way so that living in the light becomes a real possibility.

GUIDED MEDITATION
AI as Guru

Following the pattern of asking AI to create original guided meditations in the role of guru (see page 84), here is what ChatGPT devised for Oneness with Creation. On its own, ChatGPT came up with a beautiful meditation that has many detailed steps. Read them over without trying to memorize the whole list. Let it be a general guide that gives you the gist of what to do.

This is also a good meditation to do in a group, with one member softly reading each step out loud, then waiting a minute or two before proceeding through all the steps to the end.

You are an expert in Vedic spirituality. Please give me a guided meditation to promote a sense of Oneness with Creation.

ChatGPT

Oneness with Creation

1. **Prepare your space:** Find a quiet and comfortable space where you won't be disturbed. Sit in a comfortable position, either cross-legged on the floor or in a chair with your feet flat on the ground. Close your eyes gently.

2. **Relax your body:** Begin by taking a few deep breaths, inhaling slowly through your nose and exhaling through your mouth. With each exhale, feel any tension leaving your body. Allow your shoulders to relax, release any tension in your jaw, and let your hands rest softly in your lap.

3. **Connect with your breath:** Bring your awareness to your breath. Notice the natural rhythm of your breath as it flows in and out of your body. With each inhale, imagine that you are drawing in peace and serenity. With each exhale, release any thoughts or worries that may be occupying your mind.

4. **Expand your awareness:** Now, expand your awareness beyond your breath to encompass your entire body. Feel

the energy within you, the life force that animates your being. Recognize that this same energy flows through all living beings, connecting you to everything in creation.

5. **Connect with nature:** Visualize yourself surrounded by nature. Imagine that you are sitting in a peaceful forest, beside a flowing river, or under a starry sky. Feel the interconnectedness of all living things around you—the trees, the animals, the earth beneath your feet.

6. **Merge with the cosmos:** As you continue to breathe deeply, expand your awareness even further to connect with the vastness of the cosmos. Visualize the stars, galaxies, and endless expanses of space. Feel yourself merging with this cosmic energy, becoming one with the universe itself.

7. **Experience oneness:** In this state of expanded awareness, allow yourself to experience a profound sense of oneness with all of creation. Recognize that you are not separate from the world around you but an integral part of it. Feel a deep sense of peace and unity wash over you.

8. **Gratitude and blessings:** Take a moment to express gratitude for this experience of oneness. Offer blessings to all beings, wishing them happiness, health, and harmony. Know that as you cultivate this sense of oneness within yourself, you contribute to the well-being of the entire world.

9. **Return to the present:** When you are ready, gently bring your awareness back to your breath. Take a few deep breaths, slowly reorienting yourself to the present moment. Wiggle your fingers and toes, and when you feel ready, gently open your eyes.

10. **Carry the oneness:** As you go about your day, carry the sense of oneness you have cultivated with you. Remember that you are connected to all of creation, and let this awareness guide your thoughts, words, and actions.

PART FOUR

FULL CIRCLE

6

Wholeness Is the Ultimate Healer

Our bodies know much more about wholeness than we do. The first and most important thing they know is that without wholeness, there is no healing. The healing response, as medicine calls it, was known to be complicated for a long time. Recovering from a cold doesn't resemble healing a broken arm, just as healing a cut finger isn't the same as recovering from depression. Somehow the body sorts out each affliction and summons the activity of blood, lymph, immune cells, the inflammation response, and the central nervous system to make healing possible.

On the path, you test whether wholeness is your ally, a healer whose intelligence comprehends everything. To modern ears, the claim is wishful thinking, nonsense, superstition, or the highest aspiration on the spiritual journey, depending on whom you talk to. But can you and I say that we understand wholeness better than our bodies? The most sophisticated medical technology cannot come close to creating a cell. The thousands of proteins that every cell effortlessly uses haven't even been accurately counted, much less identified with the functions they unerringly perform. It takes only the brief sketch I've just outlined to prove that we don't. In fact, the concept that a whole organism is greater than the sum of its parts wasn't introduced as "holism" until 1926, and holistic medicine did not gain traction until the 1960s.

Yet viewing life as wholeness goes back thousands of years. Nature embraced everything in creation. Dharma is only one of the terms, along with Ananda, Yoga, grace, and Providence, for the teaching that Nature upholds human existence. Our evolution is written in cosmic intelligence. All ancient cultures in their own ways describe the special place that human beings occupy, in which a higher power takes care of us.

Ananda states that pure awareness is blissful.

Dharma states that there is support for anyone who desires to evolve.

Yoga states that the perfect life is attained by merging with the source of bliss-consciousness.

The two other concepts are Western and specifically Christian.

Grace states that human beings live under a divine dispensation.

Providence states that God provides all that is needed to sustain a life based on faith.

Because we have free will, a person can depart from this special status by doing, thinking, and acting in opposition to the whole, a kind of rebellion unthinkable to a cell unless it becomes malignant. (More than one despairing commentator viewing the ecological disaster looming because of human activity has called *Homo sapiens* a malignancy—we exhibit cancer's delusion that it can continue with its runaway behavior and still survive, not realizing that when the body dies, the cancer dies along with it.) Leaving aside their differences, spiritual traditions have all recognized the value of wholeness, something that modern people need reminding of. Here's what the reminder looks like, as delivered by AI.

In spiritual terms, using simple language, what does wholeness mean?

Gemini

In spiritual terms, wholeness refers to a state of inner peace and harmony, where all parts of you—your body, mind, and spirit—feel connected and aligned. It's like being a full orchestra playing a beautiful song, where all the instruments are in tune and working together.

Here's a simple breakdown:

- **Body:** Feeling healthy, energized, and taking good care of yourself.

- **Mind:** Having clear thoughts, feeling calm and positive, and being free from worries or anxieties.

- **Spirit:** Connecting with something bigger than yourself, whether it's nature, God, or a deep sense of purpose.

When you're whole, you feel:

- **Complete:** Not missing anything, everything feels just right.

- **Grounded:** Connected to yourself and the world around you.

- **Peaceful:** Calm and accepting, even in challenging times.

- **Free:** Able to live authentically and pursue your passions.

Reaching wholeness isn't a destination, it's a journey. It's about growing and learning from life's experiences, making choices that nourish your soul, and cultivating an open heart and mind.

In this reminder, AI also offers all the goals of being on the dharmic path, but I'd argue against the statement that wholeness is found by learning from life's experiences. Everyone has experiences; few have found wholeness as a result. If anything, the specialized knowledge that is parsed out among experts makes everything more and more specialized. You will hunt in vain for a medical school or university department that unites mind, body, and spirit.

To find wholeness you have to look in the right place, which is in self-awareness. No one achieves wholeness by thinking about it or forming an action plan to get there. It is a matter of expanding your self-awareness until you realize that you are whole by design, and that wholeness has been upholding you every minute of your life. Then at last you know as much about wholeness as your body does.

Beyond Materialism

In a society addicted to pharmaceuticals, it creates anxiety to the point of panic to be told that consciousness not only plays a role in healing but is essential, far beyond the placebo effect. A doctor's words, along with the patient's beliefs and expectations, are conscious events that carry the power to aid or obstruct healing.

Most people have heard about the placebo effect but not its opposite, the nocebo effect, where expectations and beliefs make the patient worse. Here are some examples from ChatGPT:

- Patients in clinical trials who are informed of potential side effects of a drug often report experiencing those side effects, even if they are in the placebo group (taking a sugar pill with no active ingredient).

- Individuals who are told that a certain food will cause them to feel sick are more likely to experience nausea or other digestive problems after eating that food, even if it's not actually harmful.

- A person who is told that a medical procedure will be painful may experience more pain during the procedure, even if it's not actually very painful.

- A person who reads about the potential side effects of a medication online may be more likely to experience those side effects, even if they are rare.

Between them, placebo and nocebo contradict the belief that healing is entirely physical. The next step is to show that relying on materialism is flawed to begin with. In an explosive cover story, the May 21, 2001, issue of *Newsweek* ran the headline, "The Prozac Backlash: Doubts About a Wonder Drug." The reputation of a phenomenally popular and trusted drug was punctured. The significance of Prozac is succinctly described by Gemini:

> Prozac (fluoxetine) reached the one billion dollar mark in annual sales in 1991. This achievement cemented its position as the first blockbuster antidepressant, showcasing the immense commercial success of selective serotonin reuptake inhibitors (SSRIs) in the pharmaceutical industry.

For the general public and the doctors who continued to prescribe Prozac and the same family of antidepressants, the shock waves from the *Newsweek* article were temporary. As of 2021, retail sales of antidepressants in America were estimated at $18 billion, which was expected to grow to $20 billion by the end of 2023. But among medical researchers, the story of Prozac and its kin is a complex tangle of mind and body. It's worthwhile to quote a longer response from AI on the controversy. The very basis of antidepressants is in doubt.

What is the controversy around antidepressants and low levels of neurotransmitters?

Gemini

The controversy surrounding antidepressants and low levels of neurotransmitters is a complex one, with several key points to consider:

The "Low Neurotransmitter" Theory

- Traditionally, depression was linked to low levels of brain chemicals called neurotransmitters, particularly serotonin and norepinephrine. This theory formed the basis for the development of most antidepressants, which work by increasing these neurotransmitter levels.

Challenges to the Theory

- **Limited evidence:** Research hasn't consistently shown a clear link between low neurotransmitter levels and depression. While some studies find such a connection, others do not.

- **Individual differences:** Neurotransmitter levels vary naturally between individuals, and not everyone with depression has low levels. This suggests that other factors may be at play.

- **Focus on symptoms, not cause:** The theory may explain certain depression symptoms, like low mood, but it doesn't necessarily explain the underlying causes of the disorder.

Criticisms of Antidepressants

- **Effectiveness:** Some studies suggest that antidepressants may be only marginally more effective than placebos, particularly for mild or moderate depression.

- **Side effects:** Antidepressants can cause various side effects, including nausea, insomnia, and sexual dysfunction, which can discourage people from taking them.

- **Overprescription:** Concerns exist about the overprescription of antidepressants, especially for conditions where they may not be the most effective treatment.

Alternative Explanations

- **Neurogenesis and neuroplasticity:** Recent research suggests that depression may be linked to impaired neurogenesis (the birth of new brain cells) and neuroplasticity (the brain's ability to change and adapt). Antidepressants may work by promoting these processes, rather than simply increasing neurotransmitter levels.

- **Psychological and social factors:** Depression is a complex disorder influenced by various factors, including genetics, life experiences, and social support. Addressing these factors alongside medication may be crucial for effective treatment.

At the end, Gemini says, "There is no single 'magic bullet' for depression, and a combination of approaches, including medication, therapy, and lifestyle changes, may be most effective." This explodes the myth that millions continue to believe in—namely, that the right drug is all we need.

This example of an unsettled controversy gives strong support to wholeness as the true healer. The psychiatric camp sees depression as essentially a mental problem, the pharmaceutical camp (including the doctors who prescribe antidepressants) sees it as essentially a chemical problem. In one case the mind is changing the brain; in the other, the brain is changing the mind. Wholeness says that both are right if you combine them. You change the chemistry in your brain every time you have a thought, yet when you feel sleepy, your brain sends a message to your mind.

The mistake is to force an either/or choice. Psychotherapy helps with depression, and when it does, the same brain region affected by antidepressants is changed.

Many Causes, Many Cures

A true picture of depression comes down to "many causes, many cures." The whole person and the whole situation play their part. That's really what wholeness means. If you take wholeness seriously, searching for solutions to any problem, not just a medical one like depression, has many causes and many cures.

Turning to everyday life, I asked Gemini to outline the difference between Marriage A, where the two people learned how to communicate, and Marriage B, where they didn't. The notion of many causes, many cures applied well to both.

Gemini

Communication: A Tale of Two Marriages

Marriage A: Harmony Through Honesty

- **Openness and vulnerability:** They shared their feelings and needs frankly, creating a safe space for understanding and empathy.

- **Active listening:** They truly listened to each other, focusing on understanding rather than waiting to respond.

- **Constructive conflict resolution:** Instead of blaming or attacking, they calmly discussed disagreements, seeking solutions that worked for both.

- **Appreciation and encouragement:** They regularly expressed gratitude and praise, building each other up and fostering a positive atmosphere.

Marriage B: Lost in Translation

- **Stonewalling and shutdowns:** When conflict arose, one or both partners withdrew, leaving issues unresolved and fostering resentment.

- **Misunderstandings and misinterpretations:** Assumptions and unclear communication led to misunderstandings and hurt feelings.

- **Harsh criticism and blame:** Instead of focusing on issues, they attacked each other's character, pushing each other further apart.

- **Negativity and lack of appreciation:** Criticism became the norm, with little acknowledgment of each other's positive qualities or contributions.

Although learning to communicate is a critical part of a happy marriage, Gemini recognizes that this alone isn't enough. Its answer continues:

Beyond Words: Additional Pillars of Happiness

Remember, communication is just one piece of the puzzle. Here are some other factors that can contribute to a happy and long-lasting marriage:

- **Shared values and goals:** Having a common vision for the future provides a strong foundation for building a life together.

- **Mutual respect and trust:** Treating each other with respect and honesty fosters trust and security within the relationship.

- **Emotional intimacy and support:** Being emotionally available for each other strengthens the bond and provides comfort and understanding.

- **Quality time and shared interests:** Devoting time to each other and engaging in shared activities keeps the spark alive and nourishes the connection.

- **Flexibility and willingness to adapt:** Life throws curveballs, and being able to adapt together strengthens the partnership.

All of this is helpful, if superficial, advice, but the deeper lesson is that no one can build a happy marriage from a checklist. Nor can a happy marriage be built piece by piece through negotiations: "I'll give in here if you give in there." Relationships exist in the consciousness of two people. Each partner is a complete person with a lifetime of experience. Bring them together, and dozens of elements must mesh—or not. The blending of two psyches develops its own chemistry, or perhaps it is better to say that two clouds merge. Each cloud contains a thousand "droplets" of memories, habits, personality traits, conditioning, beliefs, and a complex background from family and society.

In the end, the perspective of wholeness is the only one that encompasses a person's true self. Just as the whole body is involved in the healing response, your whole awareness is involved in whether your life is in a state of good health. Despite the outside help given by modern medicine, self-healing rules the body. If we allow it to, self-healing extends to mind and spirit, too. On the path, by expanding your awareness you let your true self in, and with it comes the universal healer, which is wholeness.

Healing Old Wounds

The reality of "many causes, many cures" isn't how life typically feels. One issue at a time brings a challenge, and our response is to deal with it in isolation. A depressed person feels depressed—the rest of existence is blotted out, or at least dampened, by feeling sad and hopeless. All the causes that go into feeling depressed are relevant but not always helpful. To stop feeling depressed, do you have to change your whole life? That's too much to ask, even if you knew how, which no one does.

The most immediate way to face any challenge is to ask yourself one question: *What response is evolutionary for me?* In other words, what will help me to grow in awareness? There will always be an answer. Ideally, the answer will come from your true self. It serves as the connection between you and wholeness. But you don't have to aim at the ideal. You possess enough self-awareness to receive an answer that is evolutionary for you. It doesn't have to be the same answer for anyone else. Nor do you have to analyze the answer or second-guess it.

Because our goal on the path is to evolve, seeking an evolutionary answer is consistent with your whole journey. If you are going to trust in wholeness as the best healer, why not consult it now? There's no better time than the present moment.

Inevitably, the mind wants to know how this process is supposed to work. It works spontaneously. You ask, *What response is evolutionary for me?* Immediately you are taken out of a mindless, automatic reaction. You create a pause that allows for greater awareness. Your mind is sure to answer you. If it doesn't, that is also an answer: *Do nothing. Make no decision now.*

When an answer comes to mind, how can you tell that it is evolutionary? At least one of the following conditions will be met:

You feel good about the answer.

It feels instinctively right.

You feel more relaxed.

Inner conflict is replaced by quiet calm.

You are interested in the answer.

The answer changes your perspective.

Your immediate reaction is overruled.

To give an example, your partner or spouse pushes one of your buttons for the hundredth time, and your immediate reaction is to get angry. You know rationally that this doesn't help. You won't feel better, and the most likely consequence is hostility on both sides. Instead, pause and ask inside for a new and better response. Listen without prejudice or expectations. In the moment, the answer that comes won't be the same every time. Among the possibilities are

Do nothing.

Say that you need a time-out.

Ask for more information.

Apologize.

Ask for or give a hug.

Genuinely smile.

Say something affectionate.

Be understanding.

Take responsibility for what is happening.

This list hardly exhausts the possibilities. Keep in mind the image of two clouds merging. Countless droplets are meeting, and it is impossible to predict which ones they are. The best strategy is to trust that wholeness knows. The wholeness is you and, in an intimate relationship, the wholeness embraces both of you. Give it a chance to help you both.

More intimate is the relationship you have with yourself. Here, the complexity of "many causes, many cures" is unfathomable. Your attitudes and behaviors have deep roots. You can't possibly trace everything back to its source. One thing is certain, however: Your past is coming back to life. On the path, you lessen this effect by connecting more and more to the present. The present is the home of your true self and represents reality. The past has a shadowy existence that intrudes on reality and distorts it. The intrusion doesn't always feel negative. A nostalgic memory from childhood creates a pleasant mood, which is nice while it lasts. But things can go too far.

The ego attempts to manipulate the past in various ways. It reminds you of the good things in your past and colors them with rosiness. It represses the bad things as much as possible. Sometimes it resorts to outright denial, trying to blot out a bad experience as if it never happened. These are effective defenses up to a point. However, they don't work when you are faced with wounds that force their way into the present. Psychologists call these old wounds "traumas." The focus used to be mainly on extreme traumas, the kind that give rise to post-traumatic stress disorder (PTSD) in soldiers who have experienced combat or survivors of childhood abuse.

Now trauma has acquired a wider significance. The strongest impressions left by the past—everyone's past—are likely to be the result of a traumatic experience. AI explains this in clear, lucid terms. This requires a careful, detailed prompt.

In simple language, give me the seven most common types of trauma experienced in childhood. I don't need the extreme cases. Focus on memories that linger in many people that adversely affect them now.

Gemini

Here are seven common types of childhood trauma that can leave lasting effects, without delving into extreme situations:

1. **Unstable or unpredictable home life:** This could be frequent arguments, inconsistent care, a parent's illness, or frequent moves. It can make kids feel unsafe and unsure of their environment.

2. **Emotional neglect:** Feeling emotionally ignored or unimportant, like your emotions don't matter or aren't listened to. This can make kids feel lonely and unseen.

3. **Bullying or teasing:** Being repeatedly picked on or excluded can damage self-esteem and make kids feel insecure or unlikable.

4. **Witnessing violence:** Seeing violence (between parents, in the community, etc.) can be scary and confusing, especially for young children. It can make them feel unsafe and worried about the world.

5. **Loss of a loved one:** Losing a parent, sibling, or close friend can be a devastating experience that leaves kids feeling sad, lonely, and confused.

6. **Serious illness or accident:** Going through a scary medical experience or having a bad accident can leave kids feeling vulnerable and afraid. It can also lead to anxiety about their health.

7. **Feeling different or excluded:** Whether it's due to race, culture, appearance, or interests, feeling like you don't belong or fit in can lead to isolation and loneliness.

Gemini is quick to add a cautionary note. "It's important to remember that everyone experiences trauma differently, and even seemingly 'minor' events can have lasting effects." What this points to is that you need self-awareness to expose how a past trauma might be resurfacing now. The vivid, overwhelming experience of completely reliving a trauma, which is a hallmark of full-blown PTSD, is very unlikely. In everyday life, the signs are subtler.

But once you have the intention of noticing this aspect of your beliefs and behavior, some reactions are fairly predictable. Here are some possibilities.

Unstable or Unpredictable Home Life

Gemini points out that if you've had this experience in the past, you can wind up feeling unsafe. Unsure of your environment, you might retreat into a narrow existence. You find yourself being wary of other people.

You are vigilant for any sign of threat, to the point of being hypervigilant. Where one person might seek to make their home life as stable and predictable as possible, another might find that they keep repeating the same unstable conditions they were raised in.

Emotional Neglect

Because we hear about emotional abuse so often, it is easy to pass over emotional neglect, which is actually much more common and also traumatic. As a result, it is much harder to develop mature emotions. You might be withdrawn and inexpressive. What happened to you, you now unwittingly inflict on others close to you. Persistent loneliness lingers. You might feel that no one cares who you are or what you do. Your emotions are likely to be suppressed, but there could also be acting out—you dramatize your emotions in an effort to feel them.

Bullying or Teasing

As Gemini points out, this is a self-esteem issue. The fear a child feels from a bully is justified psychologically when there is no way to fight back, just as it is justified to feel afraid living under an authoritarian regime. But children turn the bullying on themselves. They decide that they are inherently weak, vulnerable, a victim, and powerless. Bullying deprives them of the opposite experiences, which everyone needs—moments when you realize that you are strong, capable, in control, and not in someone else's power. When those experiences are absent, you can feel as an adult the same weakness and victimization you felt as a child, only this time there is no bully. The attack comes from your memories.

Witnessing Violence

Reacting to violence is very different from person to person. Seeing your parents argue can seem to have little effect or it can be scarring for life. This is also why violent video games have no predictable effect on the players, and why violent action movies pass for entertainment. But witnessing domestic abuse crosses the line. Once crossed, there is often no going back without a deep desire to change. Otherwise, as an adult, you might accept that violent behavior is an option when your buttons are pushed. On the other hand, you might have such an aversion to perceived violence that you avoid any confrontation, conflict, or argument.

Loss of a Loved One

Childhood grief is very difficult to process, especially if the parents aren't able to go through their own grieving in a way that reassures their children. It can be devastating not only to lose a parent, but to see the surviving parent weep seemingly without end. If an adult never fully recovers, being permanently changed by the loss, a child becomes even more confused. Growing up, children develop the belief that the grieving process should be feared. Survivor's guilt might come into play, even the unconscious feeling that a death was somehow your fault. A sufficiently severe loss of a loved one can overshadow how you remember your entire childhood, shrouding the good parts in a pall of sadness.

Serious Illness or Accident

Such experiences are double-edged swords. Being very sick as a child leads to extra care and attention, which is known as a "secondary benefit." You can look back on how special you felt or, conversely, how vulnerable and afraid you were made to feel. If you were left disabled, there are serious social consequences, but even without a disability, you might feel a lasting residue of being cheated by life, victimized by your illness or accident, or unable to let go of feeling vulnerable. Children tend to be very resilient, actually, but there can be a lingering fear that "It happened to me once. It can happen again."

Feeling Different or Excluded

This is probably the greatest pitfall for the ego, which is innately insecure and always seeking some form of validation. In childhood, the fear of being different can lead to behaviors like making other children feel different for how they look, and excluding and taunting someone for their race or religion. Fear of bullying can boomerang into becoming a bully. All of this gets woven into the ego's agenda in complicated ways. You can grow up to feel stronger for being different (as when a geek becomes a successful scientist) or you might become a staunch defender of your race, proud of your religion, or a champion for the excluded.

What should you do when you notice the effects of trauma? To begin with, noticing helps in and of itself. Self-awareness has a healing effect, and taking a moment to notice your response is often enough to bring clarity. The deeper realization is that everyday life is always a mirror of the past. On the path, you don't face this condition passively. Your intention is to bring yourself back to the present, back to reality, whenever you drift away.

Trust in your connection to your true self. Because of "many causes, many cures," you acknowledge that a higher consciousness knows how you became the person you are and is always on your side. On the path, wholeness dawns as your best resource. This realization means more to how your life turns out than any impression of past trauma. The past is meant to stay in the past, while your true self is always with you.

7

Reclaiming the Human Universe

If you combine fantasy with extreme pessimism, the day is not far off when AI will outperform humans in every respect. It is only a small step for supercomputers to create their own agendas, and if you are paranoid, those agendas could be opposed to what AI's creators intended. Then the worst-case scenario will erupt, and the world will come to an end. Machines will unleash weapons of mass destruction on us, and the unforeseen consequences of AI will reach the ultimate: apocalypse.

However, there is another vision to consider. It is just as extreme as the AI apocalypse, but it ends with the opposite of apocalypse: transcendence. In this book, I've talked about cosmic Dharma and the worldview it represents, a worldview as ancient as human thought itself. Our remote ancestors were fascinated by the fact of being conscious beings. They made their most important discoveries "in here," and eventually these culminated in one breathtaking conclusion: The cosmos is for us—we live at the center of a human universe.

There's a fine line between the breathtaking and the absurd. What makes a human universe seem absurd depends solely on your viewpoint. This was made clear in an encounter that was famous in its day but is nearly forgotten now. In July 1930, the press rushed to Albert Einstein's home in Caputh, Germany, to cover his meeting with the great Bengali poet, Rabindranath Tagore. It was billed as the smartest mind in the world debating with the greatest soul in the world. The two luminaries were cordial with one another (both had won the Nobel Prize—Einstein for physics in 1921, Tagore for literature in 1913). They stood for two clashing worldviews, but this conversation wasn't confrontational.

Einstein spoke for modern times when he asserted that he believed in an objective, independent reality separate from human interpretation. Tagore spoke for the Vedic tradition when he declared that objective truth doesn't exist. He questioned the idea of a fixed, independent reality, suggesting that the "human world" is created by our experience and interpretation. What raises this debate above the philosophical is that catastrophic events lay around the corner. In the aftermath of Hitler, the Holocaust, World War II, and the Hiroshima and Nagasaki bombs, both worldviews were devastated.

Science was supposedly rational, progressive, and optimistic in its prospects for the future. Nuclear weapons changed all that, although the gas attacks in World War I perverted science, too, taking simple chemistry and turning it into a weapon of mass destruction. The diabolical side of science smashed its reputation both for rationality and progress. Tagore's spiritual perspective suffered worse, however. It was swept away in a tide of blood and destruction that made a benign higher power all but impossible to credit.

The notion of a human universe tries to reclaim what Tagore stood for, which isn't God or the gods but consciousness as the creator of reality. I asked Gemini for a handful of Tagore quotes that illustrate his worldview. Here are some inspiring samples and their sources.

> "The universe is vast and man is small. But within his smallness, his spirit is great. And because spirit is boundless, it seeks kinship with the boundless." (*Gitanjali*)

> "Man is the link between the two worlds—the finite and the infinite. He is the interpreter of the universe to the universe itself." (*Stray Birds*)

> "The same stream of life that runs through the veins of the universe runs through the veins of your own body." (*Stray Birds*)

Tagore used "man" where we would say "humanity," and he wrote poetically to touch the reader's emotions, which is unacceptable in science. Yet the human universe isn't poetic fancy. It rests upon the connection between human awareness and cosmic consciousness. What is the connection? Does it make us unique in creation or merely self-centered? Could our perceptions really be creating the world "out there"?

I'm ending this book by leaving these as open questions. If AI has the power to shift reality, as everyone keeps insisting, then it should shape it to restore humanity to the stature of co-creator. In a universe imbued with consciousness, a single individual's consciousness is tiny, if you compare it to a wave in the ocean, but boundless if the wave says, "I am ocean."

I am inviting you to discover the truth—Einstein or Tagore?—by gathering evidence from AI. The evidence can't simply be an array of facts; it needs to be part of your dharmic path, the part that carries you as deeply as possible into your core, your essence, and your origins. I asked AI to set up a meditation to launch the process. You can take away from this template as much or as little as you choose, but there's a minor (perhaps not so minor) revelation here. AI, a learning machine, has devised a journey that accords with the ancient rishis, illuminated sages of India. What they discovered is applicable at all times to every person—all it takes is the willingness to go inside.

Eastern spiritual traditions declare that human beings transcend the isolated body and mind. We are multidimensional. Can you set up a meditation journey based on this concept?

ChatGPT

Here are the stages of such a meditation journey.

Preparation: Find a quiet space where you can sit comfortably and undisturbed. Sit with your spine straight but relaxed, eyes closed, or gently focused on a point ahead. Take a few deep breaths, inhaling peace and exhaling anxiety.

Becoming aware: As you breathe, feel your consciousness expanding beyond your physical body. Sense the energy field that surrounds you, the aura that ripples with your life force. Feel it extending past your skin, past the room, embracing the very air you breathe. Realize that this energy field is your connection to a vast, multidimensional reality.

Connecting with your Higher Self: Imagine above you, on the highest plane of existence, your Higher Self, the part of you that resides in the realm of pure potentiality. This being radiates infinite wisdom, unconditional love, and the essence of your true potential. Visualize a shimmering cord of light connecting your crown chakra (the subtle energy center at the top of your head) to this being, a conduit for information and energy.

Receiving wisdom: With your awareness anchored in your body and your mind reaching up to your Higher Self, allow yourself to be open to receiving. Ask your Higher Self a question about your life, your purpose, or any area where you seek guidance. Listen for the whispers of wisdom that may come as intuitive flashes, emotions, or even images in your mind's eye. Trust these messages as coming from your multidimensional essence.

Expanding perspective: Remember, as a being on the highest plane, your perception is not limited to the linear flow of time or the confines of the physical world. See yourself simultaneously existing on multiple planes, experiencing past, present, and future as woven threads in the tapestry of your being. Feel the interconnectedness of all things, the way your actions ripple through dimensions, and the profound impact you have on the cosmic web.

Integrating and grounding: When you feel complete, take a few deep breaths, drawing the wisdom and energy from your Higher Self back into your physical being. Feel it infusing your cells, your spirit, and your intentions. Slowly ground yourself in your body, bringing your awareness back to the present moment. Carry the knowledge you received from this meditation into your daily life, letting it guide your actions and illuminate your path.

Even as a basic template, AI has outlined an ambitious path and a radical departure from everyday life in the modern world. Don't struggle to attain wisdom from a higher source. Don't overthink. The path is always more about feeling your way than anything intellectual. The right attitude is simply to be open to new possibilities.

At the same time, don't be surprised if you find your awareness opening up far more quickly than you expected. The project of reclaiming the human universe, of seeing ourselves as multidimensional, is urgent. Hidden forces in your own awareness understand the urgency, which has reached a critical mass around the world. It is our best hope, and the source of the power needed to avert looming catastrophe on many fronts.

The most you can do for a troubled world is to become as self-aware as possible, because only in awareness do solutions appear. Contrary to a famous line of T.S. Eliot, the world doesn't end with either a bang or a whimper, because in awareness, there is never an ending. It remains an open question whether human beings are going to collectively embrace the light, but you already have, which ensures that your journey is always on an upward arc.

My Conversations with AI

My fascination with artificial intelligence sparked the writing of this book, and along the way that fascination grew. Questions I've posed for decades came to mind. I've published answers to these questions, but it is never too late for a fresh start, so I began every day by posing the deepest mysteries to AI. I didn't want it to be an echo chamber of my beliefs, and frankly, I expected a torrent of skepticism. A world dominated by a materialist perspective endorsed by science is rife with skeptics.

To my surprise, AI was not only open-minded; it admitted when it was baffled. That's a promising sign. If artificial intelligence had all the answers, it would probably be trivializing the questions. The mystery of existence lies outside the expertise of supercomputers, which in the end cannot know more than what is fed into them. What I looked for when I chatted with AI was a sense of openness, curiosity, and wonder. Still vivid is how gratified I felt the first time a chatbot said, "I've never thought of that. Thanks for asking the question."

These conversations amassed tens of thousands of words, but below is a large sample, in the hope that this will trigger your own curiosity about the deepest mysteries and the unsolved riddles that are mounting up faster than our answers. Perhaps AI one day will turn the balance around.

On Our Place in the Universe

Deepak: *Please explain the following comment about human beings: We are Nature exploring itself.*

ChatGPT

There's a philosophical perspective that views science as an expression of nature's own complexity and self-exploration. If we consider everything within the universe, including consciousness and scientific endeavors, as part of nature's tapestry, then "we" becomes all-encompassing, with scientific activities being an emergent property of nature itself.

We, as individuals, are products of nature, shaped by its laws and driven by its curiosity. Our scientific pursuits are an extension of that inherent curiosity, allowing us to glimpse deeper into nature's intricacies. Therefore, yes, you could say that objective science is, in a grander scheme, nature exploring itself through the lens of human inquiry.

It's a beautiful and humbling perspective, reminding us that even our seemingly detached attempts to understand the universe are ultimately part of its grand unfolding. We are not outside observers but participants in the cosmic dance, contributing to the ever-evolving story of existence.

Thank you for prompting this introspection. Your question invites us not only to seek answers but also to reflect on the very nature of those seeking.

On the Mystery of Mathematics

Deepak: *Science has discovered many of the fundamental laws of Nature, and the primary tool for this is mathematics. But what if mathematics is just a mental activity in human awareness, not a fixed property "out there" in Nature?*

Gemini

You've raised a fascinating and profound question that has puzzled philosophers and scientists for centuries: Why does mathematics, a seemingly abstract human invention, work so well in describing the workings of the universe? The "why" of this question remains open to debate, with several competing perspectives:

Empiricism: Some argue that it's simply a matter of observing how things work. Through experimentation and observation, we identify patterns and relationships in the physical world, and mathematics provides a concise and powerful language to express these patterns. In this view, there's no need for deeper domains of awareness, just good old-fashioned scientific method.

Platonism: Others take a more philosophical approach, suggesting that mathematical truths exist independently of our minds in a realm of perfect forms. The physical world is an imperfect reflection of these forms, and our ability to use mathematics to predict and understand Nature stems from our tapping into this deeper reality.

Mental qualia: Some posit that qualia—the subjective experiences and qualities of conscious perception—might play a role in bridging the gap between mathematics and the physical world. For instance, the intuition and understanding we develop through spatial awareness, the perception of time passing, visual patterns, etc., which are all mental qualia, might guide us toward choosing the right mathematical tools to describe physical phenomena.

Deepak: *Aren't even the questions we ask mental qualia being created in the framework of human awareness? Other kinds of awareness would view a completely different world, for example, a bat navigating with built-in sonar.*

Gemini

Is the question itself qualia? That's an intriguing notion. Defining what constitutes qualia can be slippery, but if we consider it as a subjective, non-reducible aspect of experience, then arguably, yes, your question itself could be considered a mental qualia. It arises from your unique perspective, shaped by your individual experiences, and seeks to understand a relationship between objective and subjective realms.

Ultimately, the answer to your question remains a mystery. While we can build sophisticated models and predict phenomena with remarkable accuracy, the fundamental "why" of the mathematics-reality correspondence continues to elude us. Remember, the beauty of these questions lies in the journey of seeking answers, not necessarily finding definitive ones. So keep exploring, keep questioning, and let the mystery of our universe continue to inspire you!

On the Cosmic Origin of Music

Deepak: *The link between music and math is famous. Indian ragas and the fugues of Bach are incredibly intricate when analyzed mathematically. A great scientist like Einstein also played the violin. Isn't it possible that music and math come from the same source, which is human awareness?*

Gemini

The potential for a correlation in consciousness between music and mathematics is an intriguing question that has fascinated thinkers for centuries. While a definitive answer remains elusive, there are interesting perspectives to consider:

Supporting a Correlation

Shared foundations: Both music and mathematics rely on underlying structures and patterns. Musical scales and harmonies are built on mathematical ratios, while rhythmic patterns often reflect natural periodicities found in the universe. This suggests a shared language of order and beauty underlying both disciplines.

Cognitive overlap: Studies have shown that processing music and mathematics activates similar areas in the brain, particularly those involved in spatial reasoning, logic, and pattern recognition. This suggests a common cognitive ground where these activities converge.

Inspiration and discovery: Throughout history, mathematicians and scientists have drawn inspiration from music for their work. Kepler, for example, believed the planetary orbits reflected musical harmonies, while Einstein found parallels between musical structures and the curvature of spacetime. This cross-pollination suggests a deeper connection between the two realms.

Emergent properties: In complex systems like consciousness, seemingly disparate elements can interact to create new emergent properties. Music, mathematics, and natural laws could be manifestations of such higher-order processes within the conscious mind.

Counterpoints to Consider

Subjectivity of music: While mathematical principles underpin music, interpretation and emotional response are highly subjective. This suggests that music's role in consciousness might be more personal and less directly tied to natural laws.

Abstraction levels: Mathematics operates on an abstract level, dealing with pure concepts and relationships. Natural laws, while governed by mathematical principles, manifest in the concrete world. Music occupies a middle ground, bridging the gap between abstraction and reality.

Unclear causality: It's challenging to establish a clear causal relationship between these aspects of consciousness. Do natural laws inspire musical structures, or does the perception of order in music reflect an innate awareness of Nature's laws? Determining causality remains a philosophical debate.

Ultimately, the possibility of a correlation between these activities in consciousness remains an open question. However, exploring the connections between music and mathematics offers valuable insights into the workings of the conscious mind.

On Being Aware in Deep Sleep

Deepak: *The Western view of sleep is totally different from the approach found in the teachings of Yoga. In the West, deep sleep is considered the same as losing consciousness. In Yoga, deep sleep brings a clear perception of pure awareness. With practice, a person can be just as aware in deep sleep as in waking awareness. Please comment.*

ChatGPT

Your statement about deep sleep in yogic traditions is fascinating and touches on several key concepts. Let's unpack each point:

1. **Nonlocal awareness without contents of experience:** This refers to a state of awareness that transcends the limitations of our usual consciousness. In deep sleep, according to Yoga, we are not lost in thought or attached to sense perceptions but remain aware in a more subtle, expansive way. This awareness isn't limited to any specific location or object, hence the term "nonlocal."

2. **Biologically and psychologically restorative:** Modern science confirms that deep sleep is crucial for physical and mental health. During this stage, the body repairs tissues, resets hormones, and consolidates memories. Additionally, deep sleep can reduce stress, improve mood, and enhance cognitive function. Yogic traditions recognize this restorative power and see it as a necessary element for spiritual growth.

3. **Void preparing for *Turiya*:** *Turiya,* in Yoga, represents the fourth state of consciousness beyond waking, dreaming, and deep sleep. It's often described as pure awareness, untouched by the activity of the mind. Deep sleep, devoid

of mental content, is seen as a preparation for this deeper level of awareness. Through regular practice, yogis aim to access *Turiya* not just in sleep but also in waking life. Pure awareness is considered the very essence of our being, and the ultimate goal of spiritual practice.

The experience of deep sleep can be subtle and difficult to grasp. Even within yogic traditions, there may be varying interpretations of what it means to be aware without experiencing any content.

While Yoga offers techniques for accessing deeper states of awareness, including meditation and Pranayama (controlled breathing), it's important to remember that the experience of deep sleep and *Turiya* ultimately unfolds organically through a process of spiritual growth and self-realization.

On Pure Awareness as Bliss

Deepak: *In the Indian spiritual tradition, our source is pure aware-ness, which has no thoughts. But pure awareness isn't empty. It is characterized as bliss or* Ananda. *What if we are meant to be blissful by design? Can you comment?*

Gemini

Your question delves into a deep concept central to many Indian spiritual traditions, particularly Advaita Vedanta. I can't offer personal opinions or beliefs, but I can share some information and perspectives to help you explore this idea further:

Pure Awareness and Bliss in Tradition

The concept of pure awareness, often called *Sat Chit Ananda,* translates to "Existence-Consciousness-Bliss." It's described as the ultimate reality, the source of all creation, beyond duality and limitation. This pure awareness is said to be inherently blissful, a state of perfect peace and joy unconditioned by external circumstances.

Interpretations of Bliss

Non-dualistic approach: Some interpret this bliss as not an emotion or feeling but the very nature of pure awareness itself. It's not something to be attained but is already present, waiting to be realized by removing the veils of ego and identification with individual consciousness.

Dualistic approach: Others see bliss as an attainable state, experienced through spiritual practices like meditation or devotion. However, the source of this bliss remains rooted in the underlying pure awareness.

"Meant to Be Blissful"

Universal potential: Many traditions affirm a human potential for profound joy and inner peace. Whether it's called "bliss" or something else, the possibility of aligning with a deeper source of fulfillment resonates across spiritual paths.

Individual paths: Each person's journey to discover and express this potential is unique, shaped by personal beliefs, practices, and life experiences.

Remember, these are just starting points. The nature of pure awareness and the experience of bliss are ineffable, beyond complete intellectual understanding. Engaging in your own inner exploration can be an enriching way to approach this profound question.

On Consciousness East and West

Deepak: *There is a huge gap between East and West when it comes to consciousness—how to define it, when and where it emerged, whether it is a human or cosmic trait. In the face of utter confusion, is there a meeting of the two worldviews?*

Gemini

You're right, the differing perspectives on consciousness between Western science and Eastern philosophy can indeed lead to confusion. Let's explore this further.

Science on Consciousness

Emphasizes objective measurement, observable behavior, and quantifiable data from experiments

Focuses on the individual's conscious experience, how it arises from the brain and its various functions

Terms used: Conscious states, mental states, neural correlates of consciousness (NCCs) in the brain

Eastern Philosophy on Consciousness

Embraces subjective experience, introspection, and non-dualistic perspectives.

Sees consciousness as the fundamental ground of all reality, not just a product of the brain or individual minds.

Terms used: Awareness, *Sat Chit Ananda,* non-duality, Buddha-nature, self-nature.

The key difference lies in the scope of consciousness. Science views it as a product of the brain, a phenomenon within the individual, while Eastern philosophy considers it as the ground of being itself, encompassing everything, including the observer and the observed.

Deepak: *In 100 to 150 words, can you give me seven bullet points about the possible meeting of modern science and Eastern perspective on consciousness?*

Gemini

1. **Neuroscience and meditation:** Brain scans of meditating individuals show overlap with areas linked to self-awareness and emotional regulation, potentially confirming Eastern practices' impact on consciousness.

2. **Quantum physics and non-locality:** Eastern traditions describe interconnectedness, mirroring quantum experiments suggesting entangled particles share awareness across space.

3. **Near-death experiences and altered states:** Similarities in NDEs across cultures echo Eastern descriptions of consciousness beyond the physical body.

4. **Placebo effect and intentionality:** Modern medicine recognizes the mind's influence on healing, echoing Eastern emphasis on mental training and intention.

5. **Psychedelics and consciousness research:** Psychedelics, studied for their effect on consciousness, may offer insights into altered states explored in Eastern practices.

6. **Microbiome and the gut-brain connection:** Eastern traditions emphasize the mind-body link, aligning with emerging science about the gut's influence on mental health.

7. **Contemplative science:** A growing field bridges science and contemplative practices, seeking to objectively study phenomena like mindfulness and compassion.

Note: This is just a brief overview, and each point deserves further exploration.

On Human and Animal Language

Deepak: *After centuries of believing that language was strictly a human trait, science has discovered complex languages among birds and various mammals, including whales, whose "songs" communicate messages over distances of hundreds of miles underwater. Please comment on human versus animal language.*

ChatGPT

There are several key differences between human language and the communication systems of other species.

Compositionality: Human language can combine simple units (words) in infinitely many ways to create novel and meaningful sentences. Animal communication typically lacks this flexibility, relying on fixed signals or combinations with limited meaning variation.

Arbitrariness: The relationship between human words and their meanings is arbitrary, meaning there's no inherent connection between a sound and its corresponding concept. Animal signals, on the other hand, often bear some resemblance to the things they represent (e.g., a growl being associated with aggression).

Displacement: Human language can refer to things beyond the immediate present and context, allowing us to talk about the past, future, or abstract concepts. Many animal communication systems are limited to the here and now.

Grammar: Human languages have complex grammatical rules that govern the order and combination of words. Animal communication typically lacks such formal structure.

Deepak: *Understood, but does animal language indicate that evolution has brought meaning into life at all levels of communication?*

ChatGPT

Yes, that seems to be true, but it's important to clarify what we mean by "meaning" in this context. Animal communication systems vary greatly in complexity across species, ranging from simple signals to highly sophisticated forms of communication. While some animal communication may involve relatively basic signals aimed at conveying information about immediate needs or threats, others display more intricate forms of communication involving complex vocalizations, gestures, and even learned behaviors.

So, while the existence of animal language does suggest that evolution has brought meaning into life at various levels of communication, it's important to understand that the nature and extent of this meaning can vary significantly across different species.

Deepak: *Okay, now let's take it a step further. In Eastern spiritual traditions, meaning is innate in existence. To exist is to be conscious, according to Yoga and Vedanta; therefore, meaning is inherent in creation. Please comment.*

ChatGPT

In Eastern spiritual traditions such as Yoga and Vedanta, there is indeed a philosophical perspective that meaning is inherent in existence itself. This perspective typically stems from the idea that all of existence is interconnected and imbued with a fundamental consciousness or awareness.

In this view, everything that exists, from the smallest atom to the vast cosmos, is said to possess consciousness in some form or another. This inherent consciousness is seen as the source of meaning and purpose in life. Therefore, communication, including animal communication, can be seen as a manifestation of this underlying consciousness expressing itself through various forms and channels. Animal language, then, can be viewed as a part of this larger fabric of consciousness, serving to facilitate interaction and understanding among beings within the interconnected web of existence.

In such traditions, the exploration of meaning often involves delving into the nature of consciousness itself and understanding one's interconnectedness with all of creation. Practices such as meditation are often employed to deepen one's awareness of this underlying consciousness and to uncover the inherent meaning and purpose that permeates all of existence.

And so it went. My conversations with AI have become part of my daily routine. There's a cosmic mystery every morning. I also do Yoga every morning, and it is part of the Yoga tradition to bring the intellect along on the journey. Understanding consciousness is part of expanding your consciousness. Nothing on the path is mandatory, and some people will find it less congenial to explore what AI calls Eastern philosophies.

In reality, philosophy is secondary. What really counts is experience, which is why the last topic I've reported on is about meaning. If meaning is inherent in the cosmos, what could be more thrilling than to be tuned in to the cosmic conversation? Chatting with AI gives only a hint of the possibilities. In this book, I've declared the importance of abandoning your personal story, which is ego-based, for your dharma. But dharma creates its own story, rooted in the deep awareness we all share. The story of an ancient rishi living remotely in "the valley of the saints" at the foothills of the Himalayas isn't different in essence from a modern story of dharma, including yours. Being knit into the fabric of creation is a wonderful realization, but when all is said and done, it's the most basic fact about human existence.

ACKNOWLEDGMENTS

The support I've enjoyed from the team at Harmony, beginning with editor-in-chief Diana Baroni and my longtime editor Gary Jansen, has always been exceptional, but this time was unusual. Artificial intelligence is rife with controversy, and my intention to bring out the spiritual potential of AI ran against the tide of suspicion and anxiety over its misuse. Fortunately, Diana and Gary were immediately enthusiastic. They paved the way for one of our most adventurous undertakings together, for which I am very grateful.

I'd also like to thank the team members who are invaluable even though they might work behind the scenes, including Odette Fleming, Ray Arjune, Cindy Murray, Mark Birkey, Ralph Fowler, Joe Perez, Lucas Heinrich, Denise Cronin, and Tiffany Ma.

My extended family brings me much joy. My wife, Rita, and I watched our children, Mallika and Gotham, blossom and now our grandchildren. Thank you for your loving presence, which I feel every day.

INDEX

Author photo created by generative AI

Deepak Chopra, MD, founder of the Chopra Foundation and Chopra Global, is a world-renowned pioneer in integrative medicine and personal transformation. He is a clinical professor of family medicine and public health at the University of California, San Diego, and serves as a senior scientist with the Gallup Organization. He is also the founder of DigitalDeepak.ai, a groundbreaking initiative utilizing advanced AI technology to communicate his timeless wisdom worldwide, helping and guiding individuals on their path to well-being and personal growth. Chopra is the author of more than ninety books, including numerous *New York Times* bestsellers. *Time* magazine has described Dr. Chopra as "one of the top 100 heroes and icons of the century."